Control Theory for Engineers

Brigitte d'Andréa-Novel
Michel De Lara

Control Theory for Engineers

A Primer

Springer

Brigitte d'Andréa-Novel
Centre de Robotique
MINES ParisTech
Paris Cedex 06
France

Michel De Lara
CERMICS
Ecole des Ponts ParisTech
Marne la Vallée Cedex 2
France

ISBN 978-3-642-44118-9 ISBN 978-3-642-34324-7 (eBook)
DOI 10.1007/978-3-642-34324-7
Springer Heidelberg New York Dordrecht London

Softcover re-print of the Hardcover 1st edition 2013

Printed on acid-free paper

Springer is part of Springer Science+Business Media (www.springer.com)

Foreword

We are surrounded by electronic devices: mobile phones, computers, traffic light regulators, etc. Many of them work automatically with inputs provided by sensors scrutinizing the environment. Plane trips contain less and less manual drive. Electrical systems are managed with automatic circuits ensuring stability. Examples abound, and our technological society will use more and more such automatic devices under the pressure of various factors. Technological innovation and costs reduction increase the penetration of so-called "smart" devices. The world increase in demand for communications technologies and for energy requires more and more coordination in a global economy. Environmental protection fosters the need for soberness in the use of resources, hence of an optimized management. Control Theory is at the heart of Information and Communication Technologies of complex systems; it can contribute to answer such challenges.

We aim at providing students and engineers with basic tools and methods to tackle dynamical systems control problems. The book is organized as an engineer classically proceeds to solve a control problem, that is elaboration of a mathematical model capturing the process behavior, analysis of this model, and design of a controller to accomplish desired objectives.

Central to Control Theory are the notions of *feedback* and of *closed-loop*. In his autobiography, *Eye of the Hurricane* (see [29]), the famous applied mathematician Richard Bellman recalls how he was led from *open-loop* solutions of control problems ("functions of time") to closed-loop solutions ("policies"): *Again the intriguing thought*: *A solution is not merely a set of functions of time*, *or a set of numbers*, *but a rule telling the decision maker what to do*; *a policy.*[1] The notion of *feedback*, as a rule mapping the state space (and the time) onto the control space, is developed throughout this text.

As with heat regulation by a thermostat, many systems are regulated by feedback and the way to elaborate such feedback controllers needs some appropriate mathematical tools which are highlighted here. This is why we are particularly

[1] To make these notions more concrete, consider traffic regulation. Fixed cycle gears for traffic lights are examples of open-loop controls, whereas closed-loop traffic lights controls adapt their timing and phasing according to current traffic conditions identified by means of sensors.

interested in systems which can be represented by a mathematical model which embodies the reactions to external inputs.

The book is divided into three Parts.

In Part I, we start by exploring a graded approach of modeling in Chap. 1. We shine the spotlight on some general principles such as mass and energy conservation laws, illustrated by many examples coming from various fields (electrical engineering, mechanics, chemistry, biological processes, etc.). In the process of elaborating a mathematical model for control purposes, we highlight particular dynamical structures and related classes of variables, like state, control, and output variables. The notion of *state* of a system is emphasized, namely a finite number of quantities which, being known at a given time instant, allow us to determine the future evolution of the system. A state is a summary of past history, sufficient for prediction. From a mathematical point of view, we focus our attention on dynamical systems whose main properties are highlighted in Chap. 2. Then, we shift our gaze from internal to external mathematical descriptions. Chapter 3 is devoted to the *frequency-domain approach* and the associated *input–output representation*, in which the system is embraced as a sort-of "black box" reacting to a set of input signals. Graphical representations are used in the case of a scalar system (one input, one output), highlighting natural definitions of *robustness* degrees such as gain or phase margins. Let us emphasize that the input–output approach is well adapted to the case of systems for which it is difficult, or sometimes impossible, to obtain mathematical dynamical models from physical laws.

Part II is devoted to system analysis to ensure stability in the neighborhood of a set point, a classical problem in control science (take off of a rocket, unstable chemical reactor. . .). An equilibrium point of a dynamical system represents a steady state of the system's model, and Chap. 4 is dedicated to their stability properties. From a practical point of view, the local stability property is expressed from a simpler model than the original one, namely the linearized or tangent model around the equilibrium point, which constitutes the first-order approximation. We will show how control laws can be elaborated from this linear model.

The difference between a *knowledge model*—devoted to better understanding, at the price of a detailed description of the "piece of reality" under scrutiny, and generally nonlinear—and a control model—simpler, often linear and used to elaborate a control law—is emphasized.[2]

---

[2] As an illustration of the differences between knowledge and control models, let us consider the climate change issue. Climatologists develop large models to improve their knowledge of the climate mechanisms under increases in greenhouse gas emissions. Such models, embodied in computer codes, integrate a huge number of physical and chemical relations. They are often obtained from discretizing partial differential equations on a planetary grid: the finer the grid, the sharper the model. The mythical 1/1 map may be seen as the ultimate form of such knowledge models. Now, climate change economists also develop models. However, the most basic of these models have a few lines and are written on a spreadsheet. Such a difference stems from different perspectives. Economists do not care to understand and embrace the physics of climate change, but they do care in how to decide about the issue: should we start reducing now (at a sure cost now, for possible, but uncertain, benefits tomorrow), or should we wait for more information and

Both *knowledge models* and *control models* have their part in Control Theory. Whereas control laws or policies are designed on the basis of a control model, they are tried and simulated on the more precise knowledge model.

To solve a local stabilization problem of a system described by a model having input variables (controls or disturbances) and output variables (measurements), a useful approach consists in computing first its tangent linearization at the desired equilibrium set point. Doing so, we reveal the model skeleton tailored for stabilization purpose. In Chap. 5, the structural *controllability and observability* properties of this tangent linear system are highlighted. The controllability property means that the system can be driven from an arbitrary state to another one by means of the control inputs. The observability property expresses that it is possible to reconstruct the entire state of the system only from the past history of partial knowledge of the state vector, that is, the output variables or measurements. We prove that a controllable and observable system can be locally asymptotically stabilized by means of a *linear feedback control law*. This study is done in Chap. 5 for continuous-time dynamical systems and in Chap. 6 for discrete-time systems. The latter, represented by difference equations, are of great importance from a practical point of view, since the digital character of computers implies that the control is fed into the system only at discrete instants. We show how to discretize continuous-time linear systems at a given sampling period. In Chaps. 5 and 6, we illuminate the links between state space and input–output representations, respectively, in continuous and discrete time.

The closed-loop system is characterized by its *stability* and *precision* properties and some sort of compromise must be achieved. The quadratic synthesis casted in Chap. 7 proposes a benchmark to display the tradeoffs, under the form of a quadratic intertemporal criterion. Moreover, we introduce a probabilistic framework to take into account perturbations and measurement errors as random variables, so that the estimation problem can be solved as a linear filtering problem whose solution is the well-known *Kalman-Bucy* filter.

At the end of Parts I and II, the local stabilization problem is solved, at least theoretically. However, some degrees of freedom are left in the control law which could be used to deal with transient behaviors. Indeed, besides stability, the control should also guarantee some robustness. In other words, the control should preserve the stability property in spite of modeling errors or unknown disturbances, such as bias on sensors or actuators, oldness of components, etc. Some answers have been given in Chap. 3 in the scalar case and will be developed in Part III.

More precisely, to take into account the effect of disturbances, the state space approach consists of adding some state variables to estimate the perturbations;

(Footnote 2 continued)
postpone costly decisions? Such a formulation is a caricature of the climate change debate, discarding details, and emphasizing other traits: economists say their models are fables. Many "fables models" have been developed to try and grasp the key economic features of the climate change issue, thus shedding light onto decision making: time preferences, discount rate, risk aversion, uncertainty, learning, etc.

however, this increases the complexity of the control law. In Chap. 8, we introduce the *polynomial approach* enabling us to obtain suitable polynomial representations which are well adapted to express the links between the outputs and the disturbances. In the multivariable case, after stabilization, degrees of freedom are left in the control law, whereas they could be used for disturbance rejection without increasing the complexity of the feedback control law.

Throughout the text, different examples are developed, both in the chapters and in the exercises. The inverted pendulum on a cart deserves special mention. This toy problem runs throughout the book because it is a prototype of a naturally unstable system that can be stabilized with the different techniques that we expose in the book.

We would like to acknowledge our institutions, Mines ParisTech and École des Ponts ParisTech, respectively, for the confidence and the liberty that they grant us. This book has benefited from valuable comments from close colleagues, with a particular mention to Jean-Philippe Chancelier. Thanks to him, we have been able to elaborate a series of computer practical works http://cermics.enpc.fr/scicoslab with the free scientific software Scicoslab that he has largely contributed to develop [16]. We feel honored that Professor Jean-Michel Coron has accepted to introduce our book, and we warmly thank him.

References

1. S. Dreyfus, Richard Bellman on the birth of dynamic programming. Oper. Res. **50**(1), 48–51 2002
2. S. Campbell, J.-P. Chancelier, R. Nikoukhah, *Modeling and Simulation in Scilab/Scicos with ScicosLab 4.4*, 2 edn. (Springer-Verlag, New York, 2010)

Preface

Control Theory is one of the most important branches of engineering science. There are already good books on this theory. But this book is an outstanding one. The authors, within less than 280 pages, successfully cover the key concepts and results of the theory. This includes, in particular, the input–output representation, Routh's criteria, the PID compensator, gain and phase margins, stability and asymptotic stability, Lyapunov functions, controllability, linear state feedback, observer, output regulator, quadratic optimization, Kalman—Bucy filter, polynomial representation, and disturbance rejection. The book deals with linear and nonlinear control systems and the control systems may be continuous or discrete in time.

I appreciate very much the excellent balance between the state space and the frequency-domain approaches. These two approaches are very important, but, in all the books I know in Control Theory, one of the two approaches is always essentially ignored. Here, one clearly sees the usefulness of each approach and how they can be used together. A special emphasis is put on modeling, which is indeed a crucial step for the application of Control Theory to real life. The proofs of the theorems and propositions are very clearly detailed. The notions and results are driven from applications. Their importance is illustrated with numerous and tutorial examples borrowed from various domains of engineering and science, as, for example, fluid mechanics, thermodynamics, classical mechanics, continuum mechanics, chemistry, electricity, and biology. The inverted pendulum on a cart, which is present throughout the whole book, allows the reader to grasp quickly the big picture on many control issues. At the end of each chapter, there are illuminating exercises. They are also issued from various domains of engineering and science. They show the power of the concepts, results, and tools described in this book.

This book is primarily intended for engineers. However, it is also very interesting for many other scientists, including, mathematicians, physicists, chemists, and biologists.

In conclusion, I cannot praise this book too much. This is a must have text. It will become soon a classical textbook in Control Theory.

Professor Jean-Michel Coron
Senior member of the Institut universitaire de France

Contents

Part I
Modelling, Dynamical Systems and Input-Output Representation

Chapter 1
Basics in Dynamical System Modelling

1.1 Introduction

In this chapter, we propose a graded approach to modelling, and we recall some principles that allow the design of mathematical models for a large class of dynamical systems. A mathematical model is a first step on the path leading to the design of a control law. In practice, a control law is an algorithm which is adapted to a mathematical representation of the system to be controlled. We intend to provide the readers with tools allowing to obtain such a representation, and we refer them to specialized literature on the subject for complements [9,48]. Due to their importance, we first describe mathematical expressions of balance equations in § 1.2, together with general principles to obtain additional phenomenological laws. Basic laws and principles of Physics are discussed in § 1.3, and applications in solid mechanics, fluid mechanics and electricity are provided in § 1.4.

Throughout the text, the time variable is designated by the letter t and, when not specified, varies in $\mathbb{R}_+$ (for continuous-time models, whereas t varies in $\mathbb{N}$ for discrete-time ones).

1.2 Balance Equations and Phenomenological Laws

In Physics, most fundamental laws and principles are formulated as mathematical equations. This is the case for the principles of mass and of energy conservation (and more generally the thermodynamics principles), the fundamental principle of dynamics and the Newton Laws in Mechanics, the Maxwell equations of Electromagnetism, etc. We come back to the mathematical expressions of these laws in § 1.3, but we stress the point that many of them often express a *balance* between quantities.

B. d'Andréa-Novel and M. De Lara, *Control Theory for Engineers*,
DOI: 10.1007/978-3-642-34324-7_1, © Springer-Verlag Berlin Heidelberg 2013

1.2.1 Balance Equations

To describe the state of a physical system at a given time t, one may use two types of quantities.

Definition 1.1 *A quantity (or variable) characterizing the local state of a system at a given time is called an* extensive quantity *(or additive quantity) if it is proportional to the volume of the local element considered.*

Example 1.2 The mass M of a body is proportional to the volume Ω of this body, and so are the internal energy $\mathfrak{E}$, the entropy $\mathfrak{H}$, and the quantity of movement $\overrightarrow{P}$. These are all extensive variables, written below as integrals with respect to the infinitesimal volume $\mathrm{d}\omega$:

$$
\begin{aligned}
M &= \int_{\Omega} \rho \, \mathrm{d}\omega \,, \\
\mathfrak{E} &= \int_{\Omega} \rho \mathfrak{e} \, \mathrm{d}\omega \,, \\
\mathfrak{H} &= \int_{\Omega} \rho \mathfrak{h} \, \mathrm{d}\omega \,, \\
\overrightarrow{P} &= \int_{\Omega} \rho \overrightarrow{v} \, \mathrm{d}\omega \,.
\end{aligned}
$$

In these expressions, ρ denotes the volumic mass of the considered element, $\mathfrak{e}$ the specific internal energy (that is, per mass unit), $\mathfrak{h}$ the specific entropy, and $\overrightarrow{v}$ the velocity vector. △

Dually, we introduce the following notion.

Definition 1.3 *A quantity (or variable) characterizing the local state of a system at a given time is called an* intensive quantity *if it is independent of the (infinitesimal) volume of the local element considered.*

Example 1.4 The volumic mass ρ, the specific internal energy $\mathfrak{e}$, the specific entropy $\mathfrak{h}$, the velocity $\bar{v}$, the pressure of a gas, and the temperature are intensive quantities. △

Now, we write the general form of the balance equation that expresses the conservation of an extensive quantity. Consider an extensive scalar quantity F written under the following integral form on a volume Ω:

$$F = \int_{\Omega} \rho f \, \mathrm{d}\omega \,.$$

Here, f is the intensive scalar variable associated with the extensive variable F, by mass unit.

If the volume Ω is limited by a surface $\partial\Omega$, assumed to be fixed, with unitary normal directed towards the outside, three terms appear in the balance equation:

- one for the variation of the quantity F by time unit t in the volume Ω,

$$\frac{\partial}{\partial t}\int_{\Omega} \rho f \, \mathrm{d}\omega ,$$

- one for the flux of the quantity F through the surface $\partial\Omega$,

$$\int_{\partial\Omega} \overrightarrow{J_F}.\overrightarrow{n} \, \mathrm{d}s ,$$

  where $\overrightarrow{J_F}$ is the unitary flux by unit of surface and of time,
- one of production of the quantity F,

$$\int_{\Omega} P_F \, \mathrm{d}\omega ,$$

  where P_F is the velocity of production of the quantity F by unit of volume.

Thus, the integral formulation of the balance equation is

$$\frac{\partial}{\partial t}\int_{\Omega} \rho f \, \mathrm{d}\omega = \int_{\Omega} P_F \, \mathrm{d}\omega - \int_{\partial\Omega} \overrightarrow{J_F}.\overrightarrow{n} \, \mathrm{d}s . \tag{1.1}$$

Under smoothness assumptions on ρ, f, $\overrightarrow{J_F}$, $\overrightarrow{n}$, the Ostrogradsky formula can be applied to (1.1), and this leads to the following *balance equation* expressed in differential form:

$$\frac{\partial}{\partial t}(\rho f) = P_F - \mathrm{div}\overrightarrow{J_F} . \tag{1.2}$$

We recall that div denotes the *divergence* (see [1] for the definition).

▷ For more details, we refer the reader to [1].

1.2.2 Phenomenological Laws

Once some variables have been fixed, the fundamental laws and principles of Physics are not sufficient to write a system of equations with as many equations as unknown variables. *Phenomenological laws* (*constitutive equations* or *constitutive relations*), appropriate for each discipline, are added to make sure that the number of equations coincides with the number of variables.

A source of phenomenological laws is the following principle: *in the neighborhood of an equilibrium, the flow of an extensive quantity is proportional to the gradient of the conjugate intensive quantity.*

The term “conjugated” is to be understood in the sense where moments and velocities are conjugate variables with respect to the Lagrangian, as in Mechanics [1]. Similarly, in Thermodynamics, energy and inverse of the temperature are conjugate variables.

Example 1.5.

$$\underset{\substack{\uparrow\\ \text{thermal flow}\\ (W.m^{-2})}}{\vec{\jmath}} \quad = \quad \overset{\substack{\text{thermal}\\ \text{conductivity}\\ \downarrow}}{-\lambda} \quad \underset{\substack{\uparrow\\ \text{temperature}}}{\nabla T} \qquad \text{(Fourier's Law)}$$

$$\underset{\substack{\uparrow\\ \text{particle flow}\\ (part.m^{-2}.s^{-1})}}{\vec{\jmath}} \quad = \quad \overset{\substack{\text{diffusibility}\\ \downarrow}}{-D} \quad \underset{\substack{\uparrow\\ \text{particle density}\\ (part.m^{-3})}}{\nabla n} \qquad \text{(Fick's Law)}$$

$$\underset{\substack{\uparrow\\ \text{electrical flow}\\ (A.m^{-2})}}{\vec{\jmath}} \quad = \quad \overset{\substack{\text{electrical}\\ \text{conductivity}\\ \downarrow}}{-\sigma} \quad \underset{\substack{\uparrow\\ \text{electrical potential}\\ (V)}}{\nabla v} \qquad \text{(Ohm's Law)}\ .$$

△

More generally, we often suppose the existence of nonlinear relations between quantities that we linearize (linear elasticity approximation, for instance). This is how many phenomenological laws are established (Ohm’s Law). In general, such laws are not necessarily linear (law of perfect gases and Van der Waals’ Law in Thermodynamics).

1.3 Basic Laws and Principles of Physics

Among the basic laws and principles of Physics, we now focus on the law of mass conservation, the principles of Thermodynamics, the point mechanics and the equations of Electromagnetism.

1.3.1 Conservation of Mass

In a general way, this law states:

$$\begin{aligned} &\text{mass variation rate of a substance in a volume} \\ = &\text{ input mass flow} \\ - &\text{ output mass flow} \\ \pm &\text{ conversion by unit of time.} \end{aligned} \tag{1.3}$$

As discussed in § 1.2.1, in the case of a substance of density $\rho(t, x)$, where $t \in \mathbb{R}$ is the time and $x \in \mathbb{R}^3$ the generic point in three-dimensional space, the conservation of mass takes the form of the following partial differential equation (see Eq. 1.2)

$$\frac{\partial \rho}{\partial t} + \mathsf{div} q = k \, , \tag{1.4}$$

where $q(t, x)$ is the flow of matter, and where the additions or substractions of mass are described by a rate $k(t, x)$ of variation of density of "sources" or of "sinks".

▷ We refer the reader to [1, 9].

1.3.2 Principles of Thermodynamics

Let us briefly recall the three principles of Thermodynamics.

The *first principle* expresses the *conservation of energy*. For any system, there exists a function U of the state of the system, the internal energy, such that, during a transformation of a state 1 towards a state 2, the sum of the work W and of the heat Q is equal to the variation of U:

$$U_2 - U_1 = W + Q \, . \tag{1.5}$$

Notice that this last equation is a balance equation.

The *second principle* expresses the *irreversibility of the evolution of macroscopic systems*. On can associate with every system in thermodynamic equilibrium two functions of the state, the absolute temperature T and the entropy S, such that

- during a quasi-static transformation, we have that $dS = \frac{\delta Q}{T}$;
- or else the transformation is irreversible and we have that $S_2 - S_1 > \int \frac{\delta Q}{T}$.

We can note that this last relation is not a balance equation.

The *third principle* expresses the *impossibility to reach the absolute zero*. The entropy of any body tends towards the same absolute zero limit.

▷ We refer the reader to [32].

1.3.3 Point Mechanics

The first *Newton's Law* (or fundamental principle of dynamics) governs the evolution of a material point submitted to forces.

Proposition 1.6 *The resultant* $\overrightarrow{F}$ *of forces applied to a material point of mass* m *and velocity* $\overrightarrow{v}$ *causes a variation of its quantity of movement* $\overrightarrow{p} = m\overrightarrow{v}$. *More precisely, we have that*

$$\overrightarrow{F} = \frac{d}{dt}(\overrightarrow{p}) = \frac{d}{dt}(m\overrightarrow{v}) = m\overrightarrow{\gamma}\ , \tag{1.6}$$

where $\overrightarrow{\gamma} = d\overrightarrow{v}/dt$ *denotes the acceleration vector of the point.*

The law of action and reaction specifies the interactions between two particles.

Proposition 1.7 *Consider two particles* 1 *and* 2 *in interaction, where* $\overrightarrow{F_{12}}$ *is the force exerted by* 1 *on* 2 *and* $\overrightarrow{F_{21}}$ *is the force exerted by* 2 *on* 1. *Then, at all times and whatever the movement of particles, we have that*

$$\overrightarrow{F_{12}} + \overrightarrow{F_{21}} = 0\ .$$

The extension of these laws to sets of points leads to new laws (conservation of the kinetic moment, etc.).

▷ We refer the reader to [32].

1.3.4 Electromagnetism Equations

There are five fundamental equations of Electromagnetism. The four *Maxwell equations* relate the density of electrical load ρ and the density of electrical flow $\overrightarrow{j}$ to the electrical field $\overrightarrow{E}$ and magnetic field $\overrightarrow{B}$ that they generate:

$$\mathrm{div}\,\overrightarrow{B} = 0 \tag{1.7a}$$

$$\mathrm{rot}\,\overrightarrow{E} = -\frac{\partial \overrightarrow{B}}{\partial t} \qquad \text{(Maxwell-Faraday)} \tag{1.7b}$$

$$\mathrm{div}\,\overrightarrow{E} = \frac{1}{\varepsilon_0}\rho \qquad \text{(Maxwell-Gauss)} \tag{1.7c}$$

$$\mathrm{rot}\,\overrightarrow{B} = \mu_0(\overrightarrow{j} + \varepsilon_0\frac{\partial \overrightarrow{E}}{\partial t}) \qquad \text{(Maxwell-Ampère).} \tag{1.7d}$$

We recall that rot denotes the *rotational* (see [1] for the definition). The quantities ε_0 and μ_0 are, respectively, the vacuum permittivity and the vacuum permeability and satisfy $\varepsilon_0\mu_0 c^2 = 1$, where c denotes the velocity of the light in vacuum.

The *Lorentz equation* describes the action exerted by the fields $\overrightarrow{E}$ and $\overrightarrow{B}$ on a particle of load q and of velocity $\overrightarrow{v}$

$$\overrightarrow{F} = q(\overrightarrow{E} + \overrightarrow{v} \wedge \overrightarrow{B}), \tag{1.8}$$

where $\wedge$ denotes the *vectorial product.*

The Maxwell-Gauss equation in (1.7c) is a form of the balance Eq. (1.2) expressing the *conservation of the electrical load.*

▷ We refer the reader to [32].

1.4 Applications in Solid Mechanics, Fluid Mechanics and Electricity

The general principles discussed above find their applications in many disciplines. We now cast a glow on illustrations in solid mechanics, fluid mechanics and electricity.

1.4.1 Solid Mechanics

We briefly recall the fundamental principles of mechanics that make it possible to express the movement of rigid bodies submitted to forces or to external torques. First, we consider the case of a body without constraints. Then, we turn the spotlight onto the case of systems submitted to nonholonomic constraints. Robotics constitutes a privileged field of applications.

1.4.1.1 Translation Movement

Translation movement is generated by the action of external *forces*. For a set of material points of a rigid body, the internal forces sum up to zero, by the principle of action and reaction recalled in Proposition 1.7. By summing the time derivatives of the quantity of movement over the set of points, we easily obtain the following result thanks to Proposition 1.6.

Proposition 1.8 *The time derivative of the quantity of movement $\overrightarrow{p}$ of a system is equal to the resultant $\overrightarrow{F}$ of all the external forces applied to the system:*

$$\overrightarrow{F} = \frac{d}{dt}(\overrightarrow{p}) \,. \tag{1.9}$$

If M is the total mass of the system and $\overrightarrow{v}_G$ is the velocity of the center of inertia G, its quantity of movement writes

$$\overrightarrow{p} = M\overrightarrow{v}_G \,.$$

One deduces the *Theorem of the center of inertia.*

Theorem 1.9 *The movement of the center of inertia G of a system is the same as that of a material point whose mass would be the total mass of the system and to which would be applied all the external forces transported in parallel at G.*

1.4.1.2 Rotation Movement

Rotation movement is generated by the action of external *torques*. Let us first recall the notions of moment of a vector with respect to a point, and the notion of kinetic moment.

Definition 1.10 *Consider a point O and a vector $\overrightarrow{P}$ pointing from a point A. The moment of the vector $\overrightarrow{P}$ with respect to the point O, denoted $\overrightarrow{\mathcal{M}}(\overrightarrow{P})_O$, is given by the vectorial product*

$$\overrightarrow{\mathcal{M}}(\overrightarrow{P})_O = \overrightarrow{OA} \wedge \overrightarrow{P} \,. \tag{1.10}$$

Definition 1.11 *Let $\overrightarrow{p} = m\overrightarrow{v}$ be the vector of quantity of movement of a material point M with mass m. The* kinetic moment *of the point M with respect to a fixed point O, denoted $\overrightarrow{\sigma_O}$, is given by:*

$$\overrightarrow{\sigma_O} = \overrightarrow{OM} \wedge \overrightarrow{p} \,. \tag{1.11}$$

The kinetic moment is the moment of the vector quantity of movement of M with respect to the fixed point O.

Let us recall now the *theorem of the kinetic moment*of a material point.

Theorem 1.12 *The moment with respect to a fixed point O of the resultant $\overrightarrow{F}$ of the forces applied to a material point is equal to the time derivative of the kinetic moment of the point with respect to the fixed point O, that is,*

$$\overrightarrow{\mathcal{M}}(\overrightarrow{F})_O = \frac{d}{dt}(\overrightarrow{\sigma_O}) \,. \tag{1.12}$$

If we consider a set of material points or a rigid body, we obtain a similar result.

Proposition 1.13 *The time derivative of the kinetic moment of a system with respect to a fixed point O is equal to the sum of the moments with respect to O of all the external forces $\overrightarrow{F}$ applied to the system:*

$$\sum \overrightarrow{\mathcal{M}}(\overrightarrow{F})_O = \frac{d}{dt}(\overrightarrow{\sigma_O}) \,. \tag{1.13}$$

Let us consider the case of a rigid body K of center of inertia G. Let $\mathcal{J}$ be a fixed frame $\{O, \overrightarrow{j_1}, \overrightarrow{j_2}, \overrightarrow{j_3}\}$ and $\mathcal{E}$ be a frame $\{G, \overrightarrow{e_1}, \overrightarrow{e_2}, \overrightarrow{e_3}\}$ attached to the body K. The frame $\mathcal{E}$ is obtained from $\mathcal{J}$ by three successive rotations: a rotation of angle θ_1 around $\overrightarrow{j_1}$, of angle θ_2 around $\overrightarrow{j_2}$, and finally of angle θ_3 around $\overrightarrow{e_3}$. Denote by $\overrightarrow{\omega}$, with components $\omega_1, \omega_2, \omega_3$ in the frame $\mathcal{E}$, the associated instantaneous rotation vector. If a resultant force $\overrightarrow{F}$ is applied to a point P of the body K, the equation of rotation (1.13) is (see [32])

$$\mathfrak{I}\dot{\overrightarrow{\omega}} + \overrightarrow{\omega} \wedge (\mathfrak{I}\overrightarrow{\omega}) = \overrightarrow{GP} \wedge \overrightarrow{F} \,, \tag{1.14}$$

where $\dot{\overrightarrow{\omega}}$ denotes the vector with components $(\dot{\omega}_1, \dot{\omega}_2, \dot{\omega}_3)$ in the base $\mathcal{E}$ (the symbol $\dot{}$ denotes the *time derivative*) and $\mathfrak{I}$ is the linear operator given in the base $\mathcal{E}$ by the *matrix of inertia* of the body K (also abusively denoted by $\mathfrak{I}$). This matrix is symmetric and is given by

$$\mathfrak{I} = \begin{pmatrix} \int_K (y_2^2 + y_3^2)\,\mathrm{d}y_1\,\mathrm{d}y_2\,\mathrm{d}y_3 & -\int_K y_1y_2\,\mathrm{d}y_1\,\mathrm{d}y_2\,\mathrm{d}y_3 & -\int_K y_1y_3\,\mathrm{d}y_1\,\mathrm{d}y_2\,\mathrm{d}y_3 \\ -\int_K y_1y_2\,\mathrm{d}y_1\,\mathrm{d}y_2\,\mathrm{d}y_3 & \int_K (y_1^2 + y_3^2)\,\mathrm{d}y_1\,\mathrm{d}y_2\,\mathrm{d}y_3 & -\int_K y_2y_3\,\mathrm{d}y_1\,\mathrm{d}y_2\,\mathrm{d}y_3 \\ -\int_K y_1y_3\,\mathrm{d}y_1\,\mathrm{d}y_2\,\mathrm{d}y_3 & -\int_K y_2y_3\,\mathrm{d}y_1\,\mathrm{d}y_2\,\mathrm{d}y_3 & \int_K (y_1^2 + y_2^2)\,\mathrm{d}y_1\,\mathrm{d}y_2\,\mathrm{d}y_3 \end{pmatrix}, \tag{1.15}$$

where the y_i are the components in the frame $\mathcal{E}$ of a generic point of K. The matrix $\mathfrak{I}$ is said to be the *principal matrix of inertia* if the frame $\mathcal{E}$ is chosen so that the matrix of inertia be diagonal, that is, if

$$\int_K y_1y_2\,\mathrm{d}y_1\,\mathrm{d}y_2\,\mathrm{d}y_3 = \int_K y_1y_3\,\mathrm{d}y_1\,\mathrm{d}y_2\,\mathrm{d}y_3 = \int_K y_2y_3\,\mathrm{d}y_1\,\mathrm{d}y_2\,\mathrm{d}y_3 = 0 \,.$$

In that case, the Eq. (1.14) becomes

$$\begin{aligned} \mathfrak{I}_{11}\dot{\omega}_1 - (\mathfrak{I}_{22} - \mathfrak{I}_{33})\omega_2\omega_3 &= \mathcal{M}_1 \\ \mathfrak{I}_{22}\dot{\omega}_2 - (\mathfrak{I}_{33} - \mathfrak{I}_{11})\omega_3\omega_1 &= \mathcal{M}_2 \\ \mathfrak{I}_{33}\dot{\omega}_3 - (\mathfrak{I}_{11} - \mathfrak{I}_{22})\omega_1\omega_2 &= \mathcal{M}_3 \,, \end{aligned} \tag{1.16}$$

where the $\mathcal{M}_i$ are the components of the vector $\overrightarrow{GP} \wedge \overrightarrow{F}$ in the frame $\mathcal{E}$. The Eq. (1.16) are known under the name of *Euler equations*.

For complex systems (robots), it can be advantageous to use the variational formalism of Euler-Lagrange to obtain the equations of movement in a systematic way (we refer the reader to [24, 37, 45, 68] for instance).

1.4.1.3 Euler-Lagrange Equation

In Lagrangian mechanics, a mechanical system is described by n generalized independent coordinates q_1, ..., q_n, called "degrees of freedom" of the system. The velocity coordinates are denoted by $\dot{q}_1$, ..., $\dot{q}_n$. Setting

$$q = (q_1, \ldots, q_n) \text{ and } \dot{q} = (\dot{q}_1, \ldots, \dot{q}_n), \tag{1.17}$$

the *Lagrangian* is

$$\mathfrak{L}(q, \dot{q}) = \mathfrak{T}(q_1, \ldots, q_n, \dot{q}_1, \ldots, \dot{q}_n) - \mathfrak{V}(q_1, \ldots, q_n), \tag{1.18}$$

where $\mathfrak{T}$ is the kinetic energy and $\mathfrak{V}$ the potential energy. The kinetic energy $\mathfrak{T}(q, \dot{q})$ is of the form

$$\mathfrak{T}(q, \dot{q}) = \frac{1}{2}\dot{q}^\top M(q)\dot{q} , \tag{1.19}$$

where $M(q)$ is an $n \times n$ symmetric positive matrix and the notation $^\top$ is for the *transpose* of a vector or of a matrix.

The *Euler-Lagrange equations* are

$$\begin{cases} \dfrac{d}{dt}\Big(\dfrac{\partial \mathfrak{L}}{\partial \dot{q}_i}\Big)\big(q(t), \dot{q}(t)\big) - \dfrac{\partial \mathfrak{L}}{\partial q_i}\big(q(t), \dot{q}(t)\big) = F_i , \quad i = 1, \ldots, n \\ \dfrac{dq}{dt}(t) = \dot{q}(t) , \end{cases} \tag{1.20}$$

where F_1, ..., F_n are the generalized forces (forces and torques) external to the system or not deriving from a potential. The Eq. (1.20) constitute a system of n differential equations of second-order in $(q, \dot{q})$ (see for example [22, 45]) of the form

$$M(q)\ddot{q} + C(q, \dot{q})\dot{q} + g(q) = F , \tag{1.21}$$

where

- $M(q)$ is the matrix of coefficients of inertial terms associated with the kinetic energy (1.19);
- $g(q)$ represents the n-vector of gravity torques, or of torques and forces deriving from a potential;
- $C(q, \dot{q})\dot{q}$ is the n-vector of centrifugal or Coriolis torques;
- F represents the n-vector of forces and/or external torques.

Remark 1.14 The matrix $C(q, \dot{q})$ can be chosen in such a way that $(\dot{M} - 2C)$ be a skew-symmetric matrix, especially if $C(q, \dot{q})$ is defined by the Christoffel symbols, so that (see for example [62]):

$$C_{kj} = \frac{1}{2}\sum_{i=1}^{n}\left\{\frac{\partial M_{kj}}{\partial q_i} + \frac{\partial M_{ki}}{\partial q_j} - \frac{\partial M_{ij}}{\partial q_k}\right\}\dot{q}_i\ . \tag{1.22}$$

◇

1.4.1.4 Systems with Nonholonomic Constraints

A mechanical system can be submitted to constraints of the form

$$\phi(q, \dot{q}) = 0\ . \tag{1.23}$$

These constraints are said to be *geometric* if they only imply position coordinates q, and *kinematic* if ϕ indeed also depends on the velocity coordinates $\dot{q}$. Generally, the kinematic constraints have a linear expression in $\dot{q}$, as

$$A(q)^\top \dot{q} = 0 \text{ with } A(q) = \big(a_1(q), \ldots, a_m(q)\big)\ , \tag{1.24}$$

where $a_1, \ldots, a_m$ are m linearly independent vectors of $\mathbb{R}^n$, so that the $n \times m$ matrix $A(q)$ is of full rank m for all q in $\mathbb{R}^n$. The number of *degrees of freedom* ($d.o.f.$) is defined as the difference between the number n of generalized coordinates and the number m of independent constraints (see for example [4, 13]):

$$d.o.f. = n - m\ . \tag{1.25}$$

Definition 1.15 *The constraints* (1.24) *are said to be* holonomic *if they are "integrable," that is, if they can be reduced to geometric constraints, and* nonholonomic *if they cannot.*

Holonomic and nonholonomic constraints are found, for example, in the phenomenon of rolling without slipping.

Let us discuss now how to take into account holonomic and nonholonomic constraints within the Euler-Lagrange formalism, to obtain the dynamical model of a mechanical system submitted to nonholonomic constraints. First, let us consider an $n \times (n - m)$ matrix $S(q) = \big(s_1(q), \ldots, s_{n-m}(q)\big)$ of full rank, and satisfying

$$A(q)^\top S(q) = 0\ . \tag{1.26}$$

By introducing the m-vector λ of Lagrange multipliers associated with the constraints, the Euler-Lagrange formalism makes it possible to obtain the dynamic behavior of a mechanical system submitted to nonholonomic constraints. With the same notations as above, we write

$$M(q)\ddot{q} + C(q, \dot{q})\dot{q} + g(q) = A(q)\lambda + B(q)F\ , \tag{1.27}$$

where $B(q)$ is an $n \times (n-m)$ matrix of full rank, under the assumption that all the degrees of freedom are directly controlled by $F_1, \ldots, F_{n-m}$. Left-multiplying (1.27) by $S(q)^\top$ and using (1.26), we eliminate the Lagrange multipliers λ, and we obtain

$$S(q)^\top [M(q)\ddot{q} + C(q,\dot{q})\dot{q} + g(q)] = S(q)^\top B(q)F \,. \tag{1.28}$$

On the other hand, the constraints (1.24) and (1.26) imply the existence of an $(n-m)$-vector $\eta(q,\dot{q})$ satisfying

$$\dot{q} = S(q)\eta \,. \tag{1.29}$$

After differentiating (1.29), we obtain:

$$\ddot{q} = S(q)\dot{\eta} + R(q,\dot{q})\eta \text{ with } R(q,\dot{q}) = \frac{dS(q)}{dq}\dot{q} \,. \tag{1.30}$$

By substitution of (1.29)–(1.30) in (1.28), we obtain the following general representation of a nonholonomic mechanical system

$$\begin{cases} W(q)\dot{\eta} = S(q)^\top (-[M(q)R(q,S(q)\eta)\eta + C(q,S(q)\eta)S(q)\eta + g(q)] + B(q)F) \\ \dot{q} = S(q)\eta \end{cases} \tag{1.31}$$

where $W(q) = S(q)^\top M(q)S(q)$ is an $(n-m)\times(n-m)$ symmetric positive matrix.

▷ We refer the reader to [4, 68, 17, 13].

1.4.2 Fluid Mechanics

We provide here some elementary principles of fluid mechanics.We recall that ∇ denotes the *gradient*, and Δ the *Laplacian* (see [1] for the definitions).

At any point x of a region of $\mathbb{R}^3$, we make the assumption that a fluid has a volumic mass $\rho(t,x)$ and a velocity $\overrightarrow{v}(t,x)$, both assumed to be smooth functions. The *conservation of mass* property, discussed in § 1.3.1, yields the so-called *continuity equation* (see 1.4):

$$\frac{\partial \rho}{\partial t} + \mathrm{div}(\rho\overrightarrow{v}) = 0 \,. \tag{1.32}$$

This single equation relating two unknown functions ρ and $\overrightarrow{v}$ is not sufficient to determine them. A vectorial equation of mechanical origin is generally added.

It is assumed that each infinitesimal element of fluid is submitted to two types of forces: *external forces* (gravitation, magnetism...) assumed to have a density $\vec{b}$, and *contact forces* exerted at the surface by the rest of the fluid. Such contact forces are usually represented under the form $\sigma \vec{n}$, where $\vec{n}$ is the normal to the surface and σ the symmetric deformations Cauchy tensor. A fluid is said to be *perfect* if the Cauchy tensor is isotropic, that is, if $\sigma = -p(t, x)I$, where I is the *identity matrix*. In that case, the *conservation of the kinetic moment* yields the *Euler equation for an ideal fluid*:

$$\frac{\partial \vec{v}}{\partial t} + (\vec{v} \cdot \nabla)\vec{v} = \vec{b} - \frac{1}{\rho}\nabla p . \tag{1.33}$$

To complete the scalar continuity equation (1.32) and the vectorial Euler equation (1.33), one generally adds energy considerations which lead to two types of fluids. A fluid is said to be *incompressible* if

$$\operatorname{div} \vec{v} = 0 .$$

A fluid is said to be *isentropic* if its internal energy admits a density w that depends on the volumic mass:

$$E_{internal} = \int \rho w(\rho)\, \mathrm{d}\omega .$$

The *conservation of energy* leads to a so-called *state equation* relating the pressure p to the volumic mass:

$$p = \rho^2 w'(\rho) .$$

To take into account the *viscous* character of a fluid, one considers that the Cauchy tensor contains a nonisotropic part called *tensor of viscosity constraints*.

1.4.2.1 The Case of Liquids

A liquid is generally assumed to be incompressible. If the initial volumic mass is assumed to be uniform ($\rho(0, x) =$ constant), one shows that ρ is constant ($\rho(t, x) =$ constant), by using the continuity equation (1.32).

A viscous incompressible fluid satisfies, in addition to the continuity equation (1.32), the *Navier-Stokes equations*

$$\begin{cases} \operatorname{div} \vec{v} = 0 \\ \dfrac{\partial \vec{v}}{\partial t} + (\vec{v} \cdot \nabla)\vec{v} + \dfrac{1}{\rho}\nabla p = \vec{b} + \mu \Delta \vec{v} , \end{cases}$$

where μ is a coefficient of viscosity.

1.4.2.2 The Case of Gases

A gas is compressible and we complete the continuity equation (1.32) and the Euler equation (1.33) by so-called "state equations" relying upon energy considerations from *Thermodynamics*.

▷ We refer the reader to [1, 20].

1.4.3 Elementary Models of Electrical Circuits

Consider an electrical circuit composed of *two-terminal electrical components* (resistances, capacities, inductances...) and of *connections* between them (electrical wires forming a network with *arcs* and *nodes*).

When the circuit has dimensions much smaller than the wavelength associated with the flow frequency, one considers that such a circuit satisfies the *Kirchoff's circuit Laws*. Each element is crossed by an *electrical intensity* (instant load flow) and submitted to an *electrical voltage* (potential from which the electrical field derives).

- Kirchhoff's current Law (or nodal rule) results from conservation of the electrical load and states that the algebraic sum of flows going in and out from a node is zero at all times.
- Kirchhoff's voltage Law (or mesh rule) results from the existence of a potential for the electrical field and states that the directed sum of the electrical potential differences around any closed network is zero at all times.

These equations are completed by the *characteristics* of the two-terminal electrical components, namely the relations between the crossing electrical intensity and the voltage at both ends.

▷ We refer the reader to [32].

1.5 Conclusion

This brief survey of basic physical laws and principles, as well as of their mathematical formulations, provides the building blocks to elaborate dynamical models discussed in the rest of the text.

Chapter 2
Finite Dimensional State-Space Models

2.1 Introduction

Throughout the text, by *model* or *system*, we mean a finite set of first-order differential equations, or of discrete-time induction equations, linking scalar quantities, distributed between *descriptive* or *internal* variables and *action* or *external* variables. This approach is restrictive, but adapted to our purposes. Formal definitions of so-called (finite dimensional) *state-models* are given in § 2.2. Several examples are detailed in § 2.3. Stationary state-models without external variables are called *dynamical systems*: they are the object of § 2.4. Linear dynamical systems form an especially important subclass that deserves its own developments in § 2.5.

In what follows, we denote by $\Re(\lambda)$ and $\Im(\lambda)$ the *real part* and the *imaginary part* of the complex number $\lambda \in \mathbb{C}$. The*complex conjugate* of the complex number $\lambda = \Re(\lambda) + i\Im(\lambda)$ is $\overline{\lambda} = \Re(\lambda) - i\Im(\lambda)$.

2.2 Definitions of State-Space Models

Some dynamical systems may be described by a finite number of quantities that allow us to determine their future evolution. Let us illustrate this with two simple examples, where we introduce the notions of *state*, *state-space*, and *state-model*.

Example 2.1 Consider the system formed of a mass m linked to a spring of stiffness k and submitted to the action of a force F as on Fig. 2.1.

If z denotes the algebraic deviation (positive or negative) of the center of gravity of the mass with respect to its equilibrium position ($z = 0$ is the abscissa of the system at rest), the fundamental principle of dynamics (1.6) yields the equation

$$m\ddot{z} = -kz + F\,. \tag{2.1}$$

B. d'Andréa-Novel and M. De Lara, *Control Theory for Engineers*,
DOI: 10.1007/978-3-642-34324-7_2, © Springer-Verlag Berlin Heidelberg 2013

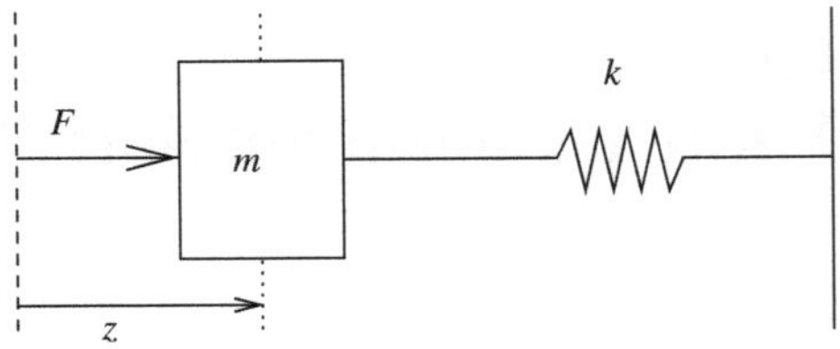

Fig. 2.1 Controlled harmonic oscillator

One can solve directly this linear differential equation with constant coefficients, giving

$$z(t) = z(t_0)\cos\big(\omega(t-t_0)\big) + \frac{\dot z(t_0)}{\omega}\sin\big(\omega(t-t_0)\big) + \frac{1}{\omega}\int_{t_0}^{t} \sin\big(\omega(t-\tau)\big)F(\tau)\,\mathrm{d}\tau,$$

where $\omega^2 = k/m$. We observe that the integration requires us to know the values of z and of $\dot z$ at a time t_0, and of $F(t)$ for $t \ge t_0$. Thus, once we know $F(t)$ for $t \ge t_0$, the evolution of $z(t)$ for $t \ge t_0$ is entirely determined by the values $z(t_0)$ and $\dot z(t_0)$. The real variables z and $\dot z$ are *sufficient* to determine the evolution of the system, once external forces are known: z and $\dot z$ are said to be *descriptive* or *internal* variables, whereas F is an *action* or *external* variable.

Although we start with the scalar equation (2.1), notice that two internal variables are necessary for its resolution. This comes from the fact that (2.1) is a *second-order* differential equation which can be written under the equivalent form of *two* differential equations of the first order:

$$\frac{d}{dt}\begin{pmatrix} z \\ \dot z \end{pmatrix} = \begin{pmatrix} \dot z \\ -\omega^2 z + \dfrac{F}{m} \end{pmatrix}. \tag{2.2}$$

The vector $x = (z, \dot z)^\top$ will be called a *state* of the previous *state-model*. △

Example 2.2 The free oscillator or pendulum without friction.

Consider a mass m suspended to a rigid thread of length l fixed at the point O, submitted to the action of the gravity field as indicated on Fig. 2.2. The fundamental principle of dynamics (1.12) for solid bodies in rotation, discussed in §1.4.1, makes it possible to write:

$$ml^2\frac{d^2\theta}{dt^2} = -mgl\sin\theta\,.$$

This equation can be put under the form

$$\frac{d}{dt}\begin{pmatrix} \theta \\ \dot\theta \end{pmatrix} = \begin{pmatrix} \dot\theta \\ -\frac{g}{l}\sin\theta \end{pmatrix}. \tag{2.3}$$

Fig. 2.2 The pendulum

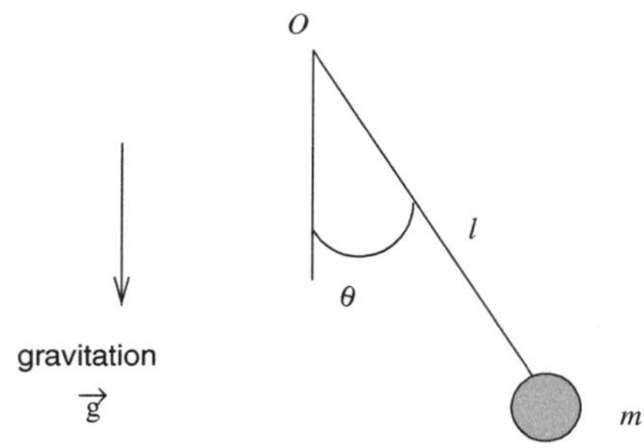

Here again, the evolution of the pendulum can be completely described by the two variables that constitute the instantaneous position θ and velocity $\dot{\theta}$ of the pendulum. Notice that θ is an angle, belonging to the unit circle S^1, so that the state $(\theta, \dot{\theta})^\top$ naturally belongs to the cylinder $S^1 \times \mathbb{R}$. △

The two examples above are special cases of the following general definition. In the sequel, by *smooth* function, we mean at least a differentiable function with continuous derivatives (of class C^1). When needed, smooth is understood as implying the required higher regularity (of class C^∞ for instance).

Definition 2.3 *Consider a system whose evolution through time $t \in \mathbb{R}_+$ is described by $n + m$ scalar quantities, denoted by $x_1(t), \dots, x_n(t), u_1(t), \dots, u_m(t)$.*

A state-model, *with* state $x = (x_1, \dots, x_n)^\top \in \mathbb{R}^n$ *and with* control $u = (u_1, \dots, u_m)^\top \in \mathbb{R}^m$ *is given by n smooth functions $f_1, \dots, f_n$ from $[0, +\infty[\times \mathbb{X} \times \mathbb{R}^m$, where $\mathbb{X}$ is an open set of $\mathbb{R}^n$, such that the evolution of $x_1(t), \dots, x_n(t)$ satisfies the following system of first-order explicit differential equations:*

$$\begin{cases} \dot{x}_1(t) = \dfrac{dx_1}{dt}(t) = f_1\big(t, x_1(t), \dots, x_n(t), u_1(t), \dots, u_m(t)\big) \\ \dots \qquad \dots \qquad \dots \\ \dot{x}_n(t) = \dfrac{dx_n}{dt}(t) = f_n\big(t, x_1(t), \dots, x_n(t), u_1(t), \dots, u_m(t)\big) . \end{cases} \tag{2.4}$$

The state-space *is the set $\mathbb{X}$ (or $\mathbb{R}^n$). This state-model is said to be* stationary *if the functions $f_1, \dots, f_n$ do not explicitly depend on the time variable t. On the contrary, we speak of a* non stationary *state-model. In what follows, we also use the terminology of (continuous-time)* state representation *to denote a state-model.*

Remark 2.4 Here, and throughout the text, the notation x has two interpretations, depending on the context. On the one hand, x can denote a *vector* of $\mathbb{R}^n$. On the other hand, x can denote a *trajectory* $t \mapsto x(t)$ from $[0, +\infty[$ towards $\mathbb{R}^n$. What we just said about the *state* x remains valid for the control u. ◇

A state-model, given by Definition 2.3, indeed makes it possible to represent the deterministic evolution of a system by the Cauchy-Lipschitz Theorem [5]. Recall that, by assumption, the dynamics components $f_1, \ldots, f_n$ in Definition 2.3 are smooth functions of their arguments.

Theorem 2.5 (Cauchy-Lipschitz) *For any initial time $t_0 \in \mathbb{R}_+$, for any initial state $\big(x_1(t_0), \ldots, x_n(t_0)\big)^\top \in \mathbb{X}$, and for all smooth mapping $t \in [t_0, +\infty) \mapsto \big(u_1(t), \ldots, u_m(t)\big)$, there exists a unique $T \in]t_0, +\infty]$ and a unique smooth solution $t \in [t_0, T) \mapsto \big(x_1(t), \ldots, x_n(t)\big)^\top$ of the system of equations* (2.4) defined on the maximal interval of time $[t_0, T[$.

The division of the $n+m$ scalar quantities in state variables $x_1(t), \ldots, x_n(t)$ and other variables $u_1(t), \ldots, u_m(t)$ corresponds to a division of the system into *internal variables*, or *state*, and *external variables*, or *inputs*. In practice, external variables can be uncontrolled perturbations as well as control variables. To bring to light such a division, we also call (2.4) a *dynamical system driven* by $u_1(t), \ldots, u_m(t)$, or a *controlled dynamical system*.

The choice of the set of variables and of the division between internal/external ones reflects the level of detail in the description of a real system. For instance, a mechanical system equipped with electrical motors can include, or not, the internal dynamical description of the motors, depending on the level of detail expected in the modelling.

Remark 2.6 The previous Definition 2.3 is contained in the more general definition of a state as a set of variables—not necessarily scalar or in finite number—which, being known at an initial time, makes it possible to determine the future evolution of the system when the external variables are known from the initial time to the current time. Thus, a state can also be a function solution of a partial differential equation (see the example of the beam in § 2.3.4), the distribution of a random vector x_t—where $(x_s)_{s\geq 0}$ is a Markov process—or a vector solution of an induction equation as in the following case of discrete-time state-models. For the control of partial differential equations, we refer the reader to [23], and for time-delay systems to [49]. ◇

Definition 2.7 *Consider a system whose evolution through time $t \in \mathbb{N}$ is described by $n+m$ scalar quantities, denoted by $x_1(t), \ldots, x_n(t), u_1(t), \ldots, u_m(t)$.*

A discrete-time state-model, *of* state $x = (x_1, \ldots, x_n)^\top \in \mathbb{R}^n$, *is given by n smooth functions $F_1, \ldots, F_n$ from $\mathbb{N} \times \mathbb{X} \times \mathbb{R}^m$, where $\mathbb{X}$ is an open set of $\mathbb{R}^n$, such that the evolution of $x_1(t), \ldots, x_n(t)$ satisfies the following system of induction equations:*

$$\begin{cases} x_1(t+1) = F_1\big(t, x_1(t), \ldots, x_n(t), u_1(t), \ldots, u_m(t)\big) \\ \quad \cdots \qquad\qquad\qquad \cdots \\ x_n(t+1) = F_n\big(t, x_1(t), \ldots, x_n(t), u_1(t), \ldots, u_m(t)\big) . \end{cases} \tag{2.5}$$

The state-space *is the set $\mathbb{X}$ (or $\mathbb{R}^n$). This state-model is said to be* stationary *if the functions $F_1, \ldots, F_n$ do not explicitly depend on the time variable t. On the contrary,* *we speak of a* nonstationary *state-model. In what follows, we use also the terminology of* discrete-time state representation *to denote a discrete-time state-model.*

In this text, we have adopted the finite dimensional state-model approach of Definitions 2.3 and 2.7. However, this is not the only way to mathematically represent the time evolution of a system (see Remark 2.6). We refer the reader to [48] for various examples of dynamical models.

2.3 Examples of Modelling

We now use several examples to illustrate how the modelling principles discussed in Chap. 1 can lead to state-models of the type discussed in § 2.2.

2.3.1 The Inverted Pendulum

The inverted pendulum with mass m on a moving cart with mass M is the mechanical system represented in Fig. 2.3. We use the *Euler-Lagrange equations* discussed in § 1.4.1 to establish a mathematical description as a state-model.

With the notations of Fig. 2.3, kinetic energy $\mathfrak{T}$ and potential energy $\mathfrak{V}$, we write

$$\mathfrak{T} = \frac{1}{2}M\dot{z}^2 + \frac{1}{2}m(\dot{z_2}^2 + \dot{y_2}^2) \text{ and } \mathfrak{V} = mgy_2 \,, \tag{2.6}$$

where, if l is the length of the rod, we have:

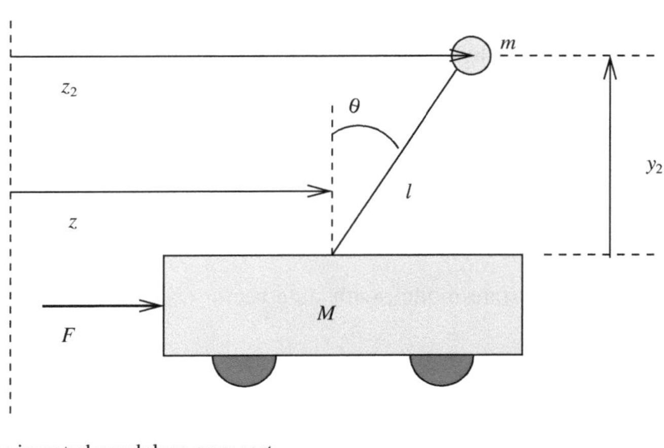

Fig. 2.3 The inverted pendulum on a cart

$$\begin{cases} z_2 = z + l \sin\theta & \dot{z_2} = \dot{z} + l\dot{\theta}\cos\theta \\ y_2 = l\cos\theta & \dot{y_2} = -l\dot{\theta}\sin\theta \ . \end{cases}$$

The Lagrangian coordinates considered are the position z of the cart and the pendulum angle θ, and the computation yields the Lagrangian $\mathfrak{L} = \mathfrak{T} - \mathfrak{V}$:

$$\mathfrak{L} = \frac{1}{2}(M+m)\dot{z}^2 + ml\dot{z}\dot{\theta}\cos\theta + \frac{1}{2}ml^2\dot{\theta}^2 - mgl\cos\theta \ . \tag{2.7}$$

We calculate

$$\begin{cases} \dfrac{\partial \mathfrak{L}}{\partial \dot{z}} = (M+m)\dot{z} + ml\dot{\theta}\cos\theta & \dfrac{\partial \mathfrak{L}}{\partial z} = 0 \\ \dfrac{\partial \mathfrak{L}}{\partial \dot{\theta}} = ml\dot{z}\cos\theta + ml^2\dot{\theta} & \dfrac{\partial \mathfrak{L}}{\partial \theta} = mgl\sin\theta - ml\dot{z}\dot{\theta}\sin\theta \ . \end{cases}$$

If F is the force exerted on the cart, the Euler-Lagrange equations (1.20) then write

$$\begin{cases} (M+m)\ddot{z} + ml\ddot{\theta}\cos\theta - ml\dot{\theta}^2\sin\theta = F, \\ ml\ddot{z}\cos\theta + ml^2\ddot{\theta} - mgl\sin\theta = 0 \ . \end{cases} \tag{2.8}$$

Now, we proceed by expressing the Eq. (2.8) under the form of a state-model with state variables $z, \theta, \dot{z}, \dot{\theta}$ and control variable, the force F. For this, we transform the two second-order differential equations (2.8), coupled and implicit, into a first-order explicit differential system.

From (2.8), we draw

$$\begin{pmatrix} M+m & ml\cos\theta \\ ml\cos\theta & ml^2 \end{pmatrix} \begin{pmatrix} \ddot{z} \\ \ddot{\theta} \end{pmatrix} = \begin{pmatrix} F + ml\dot{\theta}^2\sin\theta \\ mlg\sin\theta \end{pmatrix},$$

so that, after inverting the square matrix, we obtain

$$\begin{cases} \ddot{z} = \dfrac{F + ml\dot{\theta}^2\sin\theta - mg\cos\theta\sin\theta}{M + m\sin^2\theta} \\ \ddot{\theta} = \dfrac{-F\cos\theta - ml\dot{\theta}^2\sin\theta\cos\theta + (M+m)g\sin\theta}{l(M + m\sin^2\theta)} \ . \end{cases}$$

This gives the following state-model, with state vector $(z, \theta, \dot{z}, \dot{\theta})^\top$ and control F:

$$\frac{d}{dt}\begin{pmatrix} z \\ \theta \\ \dot{z} \\ \dot{\theta} \end{pmatrix} = \begin{pmatrix} \dot{z} \\ \dot{\theta} \\ \dfrac{ml\dot{\theta}^2 \sin\theta - mg\cos\theta\sin\theta}{M + m\sin^2\theta} \\ \dfrac{-ml\dot{\theta}^2 \sin\theta\cos\theta + (M+m)g\sin\theta}{l(M + m\sin^2\theta)} \end{pmatrix} + \frac{1}{l(M + m\sin^2\theta)} \begin{pmatrix} 0 \\ 0 \\ l \\ -\cos\theta \end{pmatrix} F. \tag{2.9}$$

▷ We refer the reader to [33, pp. 30–32].

2.3.2 A Model of Wheel on a Plane

We consider the rolling of a wheel of radius r and of mass m on a horizontal plane. Following [13, 68], we make use of the classical mechanics equations discussed in §1.4.

Consider a frame $(O, \overrightarrow{i_1}, \overrightarrow{i_2}, \overrightarrow{i_3})$ as illustrated on Fig. 2.4. We suppose that the axis of the wheel always remains horizontal. Thus, the center of mass G of the wheel moves in a plane, passing by the point O', parallel to the horizontal plane defined by the vectors $\overrightarrow{i_1}$ and $\overrightarrow{i_2}$. The vector $\overrightarrow{R} = \overrightarrow{O'G}$ can then be written

$$\overrightarrow{R} = R_1 \overrightarrow{i_1} + R_2 \overrightarrow{i_2}.$$

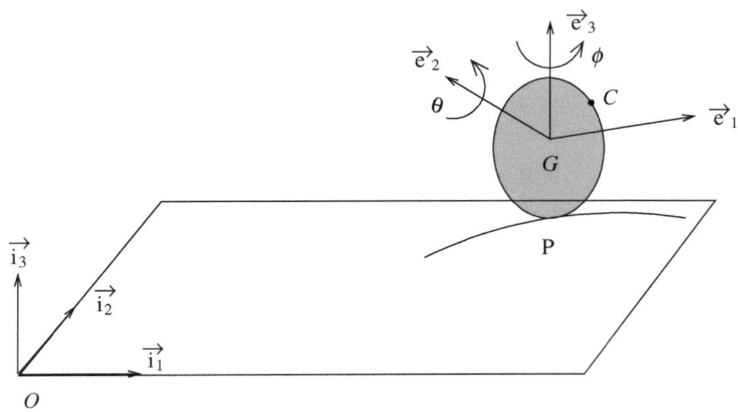

Fig. 2.4 Rolling without slipping on a plane

If we introduce an intermediary frame $(G, \overrightarrow{e_1}, \overrightarrow{e_2}, \overrightarrow{e_3})$, having vertical axis $\overrightarrow{e_3}$ and axis $\overrightarrow{e_1}$ in the direction of the movement, the passage between the frame $(O, \overrightarrow{i_1}, \overrightarrow{i_2}, \overrightarrow{i_3})$ and this intermediary frame is obtained by a rotation of angle ϕ around $\overrightarrow{i_3} = \overrightarrow{e_3}$. The passage of $(G, \overrightarrow{e_1}, \overrightarrow{e_2}, \overrightarrow{e_3})$ to a frame $(G, \overrightarrow{y_1}, \overrightarrow{y_2}, \overrightarrow{y_3})$ related to the body is obtained by a rotation of angle θ around the axis $\overrightarrow{e_2}$. As a consequence, the velocity vector $\overrightarrow{\omega}$ of instantaneous rotation of the wheel is given by

$$\overrightarrow{\omega} = \dot{\phi}\overrightarrow{e_3} + \dot{\theta}\overrightarrow{e_2}, \tag{2.10}$$

and we have the relations

$$\dot{\overrightarrow{y_i}} = \omega \wedge \overrightarrow{y_i}, \quad i = 1, 2, 3. \tag{2.11}$$

The generalized coordinates of the system are

$$q = (R_1, R_2, \theta, \phi)^\top.$$

Now, let us compute the velocity of a material point C on the periphery of the wheel, that is a point C such that

$$\overrightarrow{GC} = r\overrightarrow{y_1}.$$

The position vector $\overrightarrow{y}$ of the point C is given by

$$\overrightarrow{y} = \overrightarrow{O'C} = \overrightarrow{R} + r\overrightarrow{y_1},$$

and, by (2.11), the velocity vector of C is of the form:

$$\dot{\overrightarrow{y}} = \dot{\overrightarrow{R}} + \overrightarrow{\omega} \wedge r\overrightarrow{y_1}.$$

The property of rolling without slipping is expressed by the fact that the velocity of the material point C is zero when this point is at P, that is, when C is in contact with the base plane. In that case, $r\overrightarrow{y_1} = -r\overrightarrow{e_3}$ and we obtain:

$$\dot{\overrightarrow{y}} = 0 = \dot{\overrightarrow{R}} - \overrightarrow{\omega} \wedge r\overrightarrow{e_3} = \dot{\overrightarrow{R}} - r\dot{\theta}\overrightarrow{e_1}.$$

By decomposing this relation on the axis $\overrightarrow{e_1}$ and $\overrightarrow{e_2}$, we obtain:

$$\dot{R}_1 \cos\phi + \dot{R}_2 \sin\phi - r\dot{\theta} = 0 \text{ and } -\dot{R}_1 \sin\phi + \dot{R}_2 \cos\phi = 0\,. \tag{2.12}$$

The reader can check that these two kinematic constraints are independent and *non-holonomic*, except when the angle ϕ is constant. These constraints are expressed under the general form (1.24):

$$A(q)^\top \dot{q} = 0 \text{ with } A(q)^\top = \begin{pmatrix} \cos\phi & \sin\phi & -r & 0 \\ -\sin\phi & \cos\phi & 0 & 0 \end{pmatrix}. \tag{2.13}$$

As expected, this mechanical system has 2 degrees of freedom (see (1.25)).

To obtain a state-model of the system, we apply the *Euler-Lagrange* technique with Lagrange multipliers discussed in § 1.4.1. The 4×2 matrix

$$S(q) = \begin{pmatrix} \cos\phi & \cos\phi \\ \sin\phi & \sin\phi \\ 1/r & 1/r \\ 1/r & 0 \end{pmatrix} \tag{2.14}$$

solves (1.26), namely $A(q)^\top S(q) = 0$. By denoting $\mathfrak{J}_\theta$ (respectively $\mathfrak{J}_\phi$) the inertia of the wheel with respect to the axis $\overrightarrow{e_2}$ (respectively $\overrightarrow{e_3}$), the matrix of inertia $M(q)$, given in (1.15), is constant and diagonal, with the expression

$$M(q) = M = \begin{pmatrix} m & 0 & 0 & 0 \\ 0 & m & 0 & 0 \\ 0 & 0 & \mathfrak{J}_\theta & 0 \\ 0 & 0 & 0 & \mathfrak{J}_\phi \end{pmatrix}.$$

Notice that the nonlinear terms of coupling and of Coriolis are zero, and that the terms of gravity are exactly compensated by the reaction of the base plane. Denote by F the 2-vector of possible external torques and by $B(q)_{4\times 2}$ the matrix describing the action of the forces and external torques. By applying the Euler-Lagrange equations and the technique of multipliers, we obtain a state-model of the form (1.31)

$$\begin{cases} \dot{q} = S(q)\eta \\ W(q)\dot{\eta} = -S(q)^\top = MR(q, S(q)\eta)\eta + S(q)^\top B(q)F, \end{cases}$$

where

$$W(q) = S(q)^\top M S(q) = \begin{pmatrix} m + \dfrac{\mathfrak{J}_\theta + \mathfrak{J}_\phi}{r^2} & m + \dfrac{\mathfrak{J}_\theta}{r^2} \\ m + \dfrac{\mathfrak{J}_\theta}{r^2} & m + \dfrac{\mathfrak{J}_\theta}{r^2} \end{pmatrix}.$$

Since this last matrix is invertible, we obtain a state-model with internal variables q, η and with control F:

$$\begin{cases} \dot{q} = S(q)\eta \\ \dot{\eta} = -W(q)^{-1}\left\{S(q)^\top MR(q, S(q)\eta)\eta + S(q)^\top B(q)F\right\}. \end{cases}$$

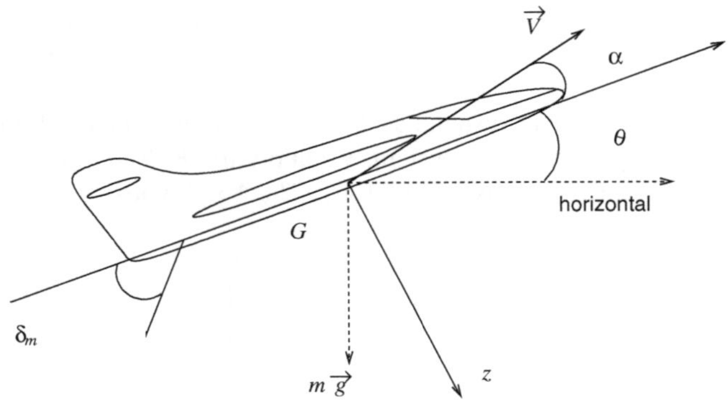

Fig. 2.5 Aircraft (longitudinal model)

2.3.3 An Aircraft Model

Consider an aircraft as in Fig. 2.5. First, let us write the equations of classical mechanics discussed in § 1.4. At the center of gravity G of the aircraft, we have

- the force equations

$$m\Gamma = F_{ae} + F_p + F_m \ , \tag{2.15}$$

- the moment equations

$$\dot{C} + \vec{\Omega} \wedge C = M_{ae} + M_p \ , \tag{2.16}$$

- the kinematic equations

$$\Gamma = \dot{\vec{V}} + \vec{\Omega} \wedge \vec{V} , \tag{2.17}$$

where

- m is the mass of the aircraft;
- Γ is the acceleration of the aircraft;
- $\vec{V}$ is the translation velocity vector of the aircraft;
- $\vec{\Omega}$ is the rotation velocity vector of the aircraft with respect to the center of gravity G;
- C is the kinetic angular moment defined by $C = \mathfrak{I}\vec{\Omega}$, where $\mathfrak{I}$ denotes the matrix of inertia;
- F_m is the vector of mass forces (weight);
- F_{ae} and M_{ae} are the vectors of aerodynamical forces and moments;
- F_p and M_p are the vectors of propulsion forces and moments.

By convention, the previous equations are written according to their projections on the axis linked to the aircraft. The orientation of these axis with respect to a fixed frame is defined by the Euler angles θ, φ and ψ, called yaw, pitch and roll, respectively. The components of the rotation vector $\overrightarrow{\Omega}$ are then denoted p, q and r, and those of the velocity vector $\overrightarrow{V}$ are denoted u, v and w.

Now, we use phenomenological laws, as discussed in § 1.2.2, appropriate for Aerodynamics. The aerodynamical forces and moments can be expressed by means of dimensionless vectors $C_F = (C_x, C_y, C_z)^\top$ and $C_M = (C_l, C_m, C_n)^\top$ as follows

$$F_{ae} = -\frac{1}{2}\rho(z)SV^2C_F \quad and \quad M_{ae} = -\frac{1}{2}\rho(z)SLV^2C_M ,$$

where

- ρ denotes the air volumic mass which depends on the altitude z;
- S is the reference surface of the aircraft;
- L is the reference length of the aircraft.

The aerodynamical coefficients C_x, C_y, C_z, C_l, C_m, C_n are complicated functions of the Mach number, and of different angles α, β, p, q, r, δ_m, δ_l and δ_n, that we do not detail. We refer the reader to [65] for expressions of these aerodynamical coefficients. In practice, the deflections δ_m, δ_l and δ_n are control variables. Consider now the acceleration vector J, as measured on the aircraft, with components J_x, J_y, J_z:

$$J = \Gamma - \frac{F_m}{m} .$$

If we suppose that the resultant of the forces of propulsion is in the symmetry plane of the aircraft, following the longitudinal axis, we obtain:

$$\begin{cases} F_{p_y} = F_{p_z} = 0 \\ M_{p_x} = M_{p_y} = M_{p_z} = 0 \\ v = 0 \\ \beta = 0 \\ p = r = 0 \\ C_y = 0 \\ C_l = C_n = 0 \\ \delta_l = \delta_n = 0 \\ \varphi = 0 . \end{cases}$$

At last, again by the assumption of longitudinal movements, the matrix of inertia has the expression:

$$\mathfrak{I} = \begin{bmatrix} \mathfrak{I}_{xx} & 0 & \mathfrak{I}_{xz} \\ 0 & \mathfrak{I}_{yy} & 0 \\ \mathfrak{I}_{xz} & 0 & \mathfrak{I}_{zz} \end{bmatrix} .$$

By introducing the constant $k = \rho SV^2/2$, the Eqs. (2.15), (2.16) and (2.17) then become (in projection on the axis Gx and Gz):

$$\begin{cases} mJ_x = -kC_x + F_{p_x} \\ mJ_z = -kC_z \\ \mathfrak{I}_{yy}\dot{q} = kLC_m \\ \dot{u} = -qw - g\sin\theta + J_x \\ \dot{w} = qu + g\cos\theta + J_z \ . \end{cases} \tag{2.18}$$

On the other hand, we have the relations:

$$\begin{cases} \dot{\theta} = q \\ V^2 = u^2 + w^2 \\ \tan\alpha = \dfrac{w}{u} \ . \end{cases}$$

The reader may verify that a state-model may be deduced of the above equations. When the movements are small around of a flying point, one may linearize the equations (2.18) in the neighborhood of equilibrium values.

▷ We refer the reader to [33, 65].

2.3.4 Vibrations of a Beam

Here is an example of a model that is not a state-model, in the restricted sense of Definition 2.3, because it involves an infinite dimensional state satisfying a partial differential equation.

We study the behavior of a beam, simply supported at one extremity and controlled at the other by a point force and a torque. We suppose that the internal forces do not work, so that, in absence of external forces, energy is conserved (see § 1.3.2).

We make the assumption of small displacements in the plane (y, z) represented in Fig. 2.6. The curve $z = z(t, y)$ represents, at each time t, the elastic axis of the beam in flexion. The length of the beam is l, the density is $\rho(y)$ and the energy at time t is $E(t)$.

The Euler-Bernoulli model is often used for its simplicity. It relies on the assumption that each infinitesimal element of beam is rigid, with a rectangular section, so that the energy can be written as

$$E(t) = \frac{1}{2}\int_0^l \left[\rho\left(\frac{\partial z}{\partial t}\right)^2 + R_f\left(\frac{\partial^2 z}{\partial y^2}\right)^2\right] \mathrm{d}y \ , \tag{2.19}$$

where the function $R_f(y)$ is the so-called "rigidity in flexion." Under the assumption of energy conservation, we obtain the equation of evolution of Euler-Bernoulli by differentiation with respect to time t and by two integrations by parts of (2.19):

$$\rho\frac{\partial^2 z}{\partial t^2} + \frac{\partial^2}{\partial y^2}\left(R_f\frac{\partial^2 z}{\partial y^2}\right) = 0 \ . \tag{2.20}$$

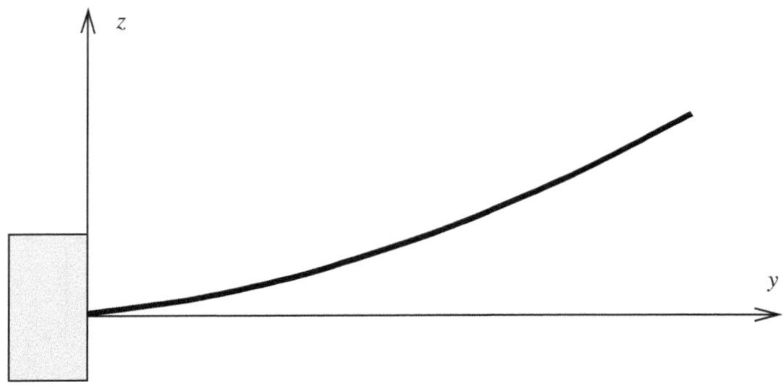

Fig. 2.6 Simply supported beam

The boundary condition expressing the support at the extremity $y = 0$ gives

$$z(0, t) = \frac{\partial z}{\partial y}(0, t) = 0 \,. \tag{2.21}$$

If $F(t)$ and $M(t)$ denote the lateral force and the moment of flexion, respectively, applied to the extremity l, we have that:

$$\begin{cases} R_f(l)\dfrac{\partial^2 z}{\partial y^2}(l, t) = M(t) \\ -\dfrac{\partial}{\partial y}\left(R_f \dfrac{\partial^2 z}{\partial y^2}\right)_{|y=l} = F(t) \,. \end{cases} \tag{2.22}$$

The equations (2.20), (2.21) and (2.22) constitute a dynamical model of the beam.

▷ For more information on beam models, we refer the reader to [59].

2.3.5 An RLC Electrical Circuit

Consider an *RLC* series electrical circuit formed of a resistor, with electrical resistance R, of a capacitor (condenser), with capacitance C, and of an inductor, with inductance L, submitted to a voltage v as in Fig. 2.7. We suppose that the *Kirchoff's Laws* discussed in § 1.4.3 apply here.

By the *nodal rule*, an electrical current, denoted i, passes through the three two-terminal electrical components of the circuit. On the other hand, these components are assumed to follow the phenomenological laws

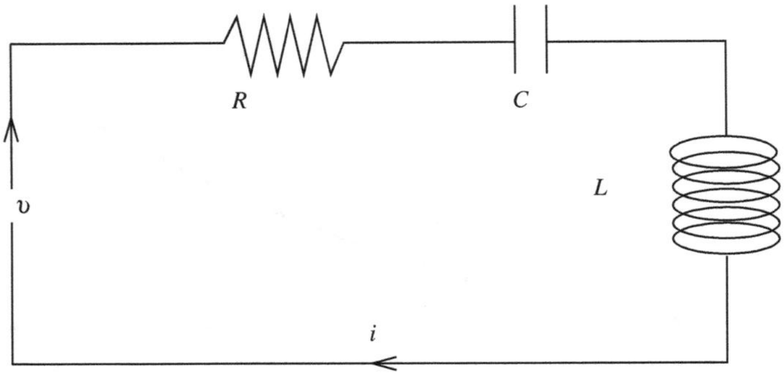

Fig. 2.7 RLC electrical circuit

$$\begin{cases} \text{resistor terminal voltage} &= Ri \quad \text{(Ohm's Law)} \\ \text{capacitor terminal voltage} &= \dfrac{q}{C} \\ \text{inductor terminal voltage} &= L\dfrac{di}{dt} \end{cases} \tag{2.23}$$

where $q(t) = \int_0^t i(s)\,\mathrm{d}s$ denotes the electrical load of the capacity.

By the *mesh rule*, the electrical voltages add up and we obtain a dynamical equation:

$$L\frac{di}{dt} + Ri + \frac{1}{C}q = \upsilon \,. \tag{2.24}$$

This last equation is equivalent to the following state-model

$$\frac{d}{dt}\begin{pmatrix} q \\ i \end{pmatrix} = \begin{pmatrix} i \\ -\frac{R}{L}i - \frac{1}{LC}q + \frac{1}{L}\upsilon \end{pmatrix}, \tag{2.25}$$

with internal variables q, i and with external variable υ. Notice that this model is linear in all the variables.

2.3.6 An Electrical Motor

An electrical motor consists of several coils of narrow thickness having a common diameter through which passes the axis of the motor. It is embedded in a uniform magnetic field orthogonal to this axis, and is supplied by a flow of intensity i under a voltage υ.

The *Ampère's Law* is a consequence of the *Maxwell equations* discussed in § 1.3.4. Ampère's Law makes it possible to compute the force exerted by the magnetic field in each coil, and thus the total torque C. This torque is proportional to the intensity i, with a coefficient almost independent of the angles between the coils and the field if there are enough coils:

$$C = K_1 i \ .$$

If the motor entails a mass of moment of inertia J, then this torque transmits to the mass an angular velocity ω which, by *conservation of the kinetic moment*, satisfies

$$J\frac{d\omega}{dt} = C \ .$$

But, by the *Lenz's Law*, the rotation of the coils in the magnetic field has the effect to develop an opposed electromotive force e. This force is proportional to the variation of magnetic flux through the coils, hence to the angular velocity ω, with a coefficient almost independent of the angles between the coils and the field:

$$e = K_2 \omega \ .$$

At last, if R denotes the electrical resistance of the motor, the *Ohm's Law* yields the relation

$$\upsilon - e = Ri \ .$$

These equations lead to a dynamical relation between the applied voltage υ and the angular velocity ω

$$\frac{d\omega}{dt} = -\frac{K_1 K_2}{JR}\omega + \frac{K_1}{JR}\upsilon \ . \tag{2.26}$$

This is a state-model, with internal variable ω and external variable υ.

▷ We refer the reader to [33].

2.3.7 Chemical Kinetics

Consider a reaction between n reactants $A_1, \ldots, A_i, \ldots, A_n$, with *stoichiometric coefficients* $\nu_{A,1}, \ldots, \nu_{A,i}, \ldots, \nu_{A,n}$, and forming p products $P_1, \ldots, P_i, \ldots, P_p$ with stoichiometric coefficients $\nu_{P,1}, \ldots, \nu_{P,i}, \ldots, \nu_{P,p}$:

$$\nu_{A,1}A_1 + \ldots + \nu_{A,i}A_i + \ldots + \nu_{A,n}A_n \underset{k_{-1}}{\overset{k_1}{\rightleftharpoons}} \nu_{P,1}P_1 + \ldots + \nu_{P,i}P_i + \ldots + \nu_{P,p}P_p \ . \tag{2.27}$$

The *reaction speed* v is defined as the speed of disappearance of a reactant or the speed of apparition of a product, taking into account the stoichiometric coefficients. Indeed, by *mass conservation*, as discussed in § 1.3.1, the chemical reaction (2.27) yields the following $n + p - 1$ equations satisfied by the *concentrations* $[A_1], \ldots, [A_n], [P_1], \ldots, [P_p]$:

$$v = -\frac{1}{\nu_{A,i}}\frac{d[A_i]}{dt} = \frac{1}{\nu_{P,j}}\frac{d[P_j]}{dt}, \quad i = 1, \ldots, n, \; j = 1, \ldots, p \,. \tag{2.28}$$

The *chemical kinetic equation* relates the reaction speed v to the concentrations $[A_1], \ldots, [A_n], [P_1], \ldots, [P_p]$, under the general form:

$$v = -k_1 \prod_{i=1}^{n} [A_i]^{\gamma_{A,i}} + k_{-1} \prod_{j=1}^{p} [P_j]^{\gamma_{P,j}} \,. \tag{2.29}$$

The partial order of the reaction is the exponent by which a component is raised in the formulation of the speed, which is not necessarily an integer. The *order of the reaction* is defined as the sum of the partial orders of the reaction with respect to each reactant or product.

Such equations provide a state-model. For example, for the reaction

$$CO + \frac{1}{2}O_2 \underset{k_{-1}}{\overset{k_1}{\rightleftharpoons}} CO_2 \,,$$

an expression of the kinetics is

$$v = \frac{d[CO_2]}{dt} = k_1[CO][O_2]^{\frac{1}{2}} - k_{-1}[CO_2] \,.$$

With $x = ([CO], [O_2], [CO_2])^\top$ for state, this gives the model:

$$\begin{cases} \dfrac{d[CO]}{dt} = -k_1[CO][O_2]^{\frac{1}{2}} + k_{-1}[CO_2] \\ \dfrac{d[O_2]}{dt} = -\dfrac{1}{2}k_1[CO][O_2]^{\frac{1}{2}} + \dfrac{1}{2}k_{-1}[CO_2] \\ \dfrac{d[CO_2]}{dt} = k_1[CO][O_2]^{\frac{1}{2}} - k_{-1}[CO_2] \,. \end{cases} \tag{2.30}$$

Notice that, with $Z = ([CO]+[CO_2], [CO_2]+2[O_2], [CO_2])^\top = (Z_1, Z_2, Z_3)$ for state, we obtain another state-model, in which the relations of conservation appear more clearly:

$$\begin{cases} \dfrac{dZ_1}{dt} = 0 \\ \dfrac{dZ_2}{dt} = 0 \\ \dfrac{dZ_3}{dt} = \dfrac{1}{\sqrt{2}} k_1 (Z_1 - Z_3)(Z_2 - Z_3)^{\frac{1}{2}} - k_{-1} Z_3 \,. \end{cases} \tag{2.31}$$

These state-models do not include external variables, but one can represent the influence of the external temperature T under the form $k_1 = k_1(T)$, $k_{-1} = k_{-1}(T)$.

▷ We refer the reader to [7].

2.3.8 *Growth of an Age-Structured Population*

Consider a population subdivided in n age-classes. This may be an animal population (birds, fishes), a vegetal population (trees), or a "population" of cars classified according to their registration year. We design a generic state-model in discrete-time as in Definition 2.7.

Let $N(t) = \big(N_1(t), \dots, N_n(t)\big)^\top$ denote the vector of abundances in each age-class at time t, where $t \in \mathbb{N}$ is discrete and any period $[t, t+1]$ coincides with that used to define a class: $N_1(t)$ is the number of individuals of (strictly) less than 1 year of age, $N_2(t)$ is the number of individuals of 1 year, ..., $N_{n-1}(t)$ is the number of individuals of $n-2$ years of age, and $N_n(t)$ is the number of individuals of more than $n-1$ years of age.

The *conservation of the number of individuals* is expressed in all generality under the form

$$N_i(t+1) - N_i(t) = B_i([t,t+1]) - D_i([t,t+1]) + I_i([t,t+1]) - E_i([t,t+1]) \tag{2.32}$$

where, on the time interval $[t, t+1]$ and for the age-class i,

- $B_i([t, t+1])$ denotes the number of births,
- $D_i([t, t+1])$ the number of deaths,
- $I_i([t, t+1])$ the number of immigrants,
- $E_i([t, t+1])$ the number of emigrants.

These fairly general equations (2.32) for $i = 1, \dots, n$ are completed with others, according to additional assumptions on the population growth. For instance, suppose that, on the time interval $[t, t+1]$,

1. a fraction of the $N_i(t)$ individuals of each class of age i disappears (death process), and the remaining fraction q_i survives and reaches age i, thus filling $N_{i+1}(t)$ (except possibly for the last class);
2. individuals appear in the first age-class (birth process), where each age-class contributes proportionally to its abundance, with fertility coefficient f_i.

Under these assumptions, one may write the following evolution model:

$$N(t+1) = \begin{pmatrix} f_1 & f_2 & f_3 & \dots & f_{n-1} & f_n \\ q_1 & 0 & 0 & \dots & \dots & \dots \\ 0 & q_2 & 0 & 0 & \dots & \dots \\ \dots & \dots & \dots & \dots & \dots & \dots \\ \dots & \dots & \dots & 0 & q_{n-1} & 0 \end{pmatrix} N(t) \ . \tag{2.33}$$

▷ We refer the reader to [19, 27].

2.3.9 A Bioreactor

Consider a bioreactor as in Fig. 2.8. It is constituted of a vessel where biomass (living cells) and nutriment interact. Moreover, a flow D brings in nutriment and emits the products.

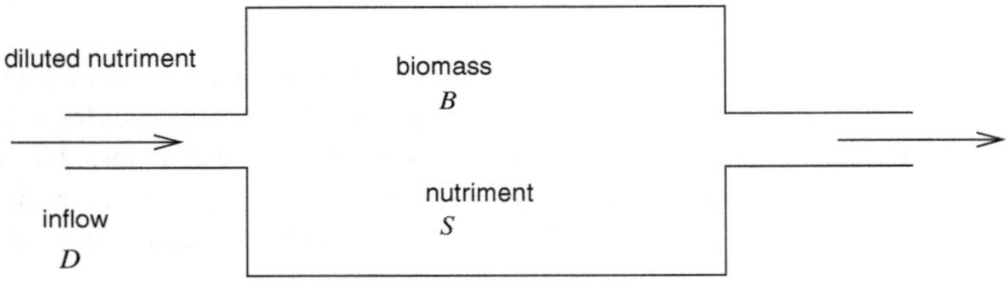

Fig. 2.8 A bioreactor

To write *mass balance equations* of the form (1.3), we introduce the following variables:

- B: concentration of biomass in the bioreactor (gl^{-1});
- S: concentration of nutriment in the bioreactor (gl^{-1});
- D: inflow of nutriment (lS^{-1});
- S_{in}: concentration of nutriment before dilution (gl^{-1}).

The balance equations are

$$\begin{cases} \dfrac{dB}{dt} = \text{growth} - DB \\ \dfrac{dS}{dt} = \text{consumption} - DS + DS_{in} \ . \end{cases}$$

To complete these equations, we introduce *phenomenological laws*, expressing the conversion between biomass and nutriment, of the form

$$\text{growth} \propto \text{biomass and consumption} \propto \text{biomass},$$

where the proportionality coefficients only depend on the nutriment:

- $\mu(S)$: specific growth rate of the biomass (S^{-1});
- $\nu(S)$: specific consumption rate of the nutriment (S^{-1}).

The combination of the balance equations with the conversion equations yields the following state-model:

$$\begin{cases} \dfrac{dB}{dt} = \mu(S)B - DB & = \text{growth} - \text{output} \\ \dfrac{dS}{dt} = -\nu(S)B - DS + DS_{in} & = \text{consumption} - \text{output} \\ & \quad + \text{feed-in}. \end{cases} \tag{2.34}$$

Generally, the functions μ and ν are assumed to be increasing with zero value at $S = 0$. The flow D may be considered as a control variable.

Remark 2.8 We have described a simple model with one type of cell and one nutriment. However, the above formulation may be extended to other situations like, for example, in a sewage treatment plant (with a higher number of state variables). ◇

▷ We refer the reader to [8].

2.4 Dynamical Systems

We now develop theoretical notions about a central class of state models, the so-called *dynamical systems*.

Definition 2.9 *We call* free dynamical system *a state-model without external variables and* (classical) dynamical system *a stationary state-model without external variables.*

A stationary state-model where the external variables are either constant or are smooth maps of the state (closed-loop) yields a dynamical system. This abstract concept of dynamical system will be useful for the study of the closed-loop stability discussed in Chap. 4. Let us introduce the following definitions.

Definition 2.10 *A* vector field f *on an open set* $\mathbb{X}$ *of* $\mathbb{R}^n$ *is a smooth mapping* $f : \mathbb{X} \to \mathbb{R}^n$ *that associates with each point* x *of* $\mathbb{X}$ *a vector* $f(x) \in \mathbb{R}^n$ *having this point for origin.*

A dynamical system *on* $\mathbb{X}$ *is a couple* $(\mathbb{X}, f)$ *where* f *is a vector field on* $\mathbb{X}$*. The open set* $\mathbb{X}$ *is called* phase space *of the dynamical system.*

Remark 2.11 More generally, on a manifold $\mathbb{X}$ of class C^∞, we mean by vector field f a vector field of class C^∞ [10]. In the system of coordinates $x_1, \dots, x_n$, the field is defined by n C^∞ functions $f_1(x_1, \dots, x_n), \dots, f_n(x_1, \dots, x_n)$. In the cases discussed in § 2.2, $\mathbb{X} = \mathbb{R}^2$ for the mass-spring system, whereas, for the oscillator, $\mathbb{X}$ is the cylinder $S^1 \times \mathbb{R}$. ◇

Definition 2.12 *We call* integral curves *of the vector field f on the open set $\mathbb{X}$ the solutions of the differential equation*

$$\dot{x} = f(x) , \tag{2.35}$$

that is, the curves $\varphi : I \to \mathbb{X}$, where $I \subset \mathbb{R}$ is an interval, such that:

$$\frac{d\varphi}{dt}(t) = f\big(\varphi(t)\big) , \quad \forall t \in I . \tag{2.36}$$

An integral curve is said to be maximal if it is not contained in any other integral curve.

If, in the system of coordinates $x_1, \ldots, x_n$, the vector field f is defined by n smooth functions $f_1(x_1, \ldots, x_n), \ldots, f_n(x_1, \ldots, x_n)$, then the differential equation (2.35) may be written under the form of the following differential system with solution $x_1(t), \ldots, x_n(t)$

$$\begin{cases} \dot{x}_1(t) = f_1\big(x_1(t), \ldots, x_n(t)\big) \\ \cdots = \cdots \\ \dot{x}_n(t) = f_n\big(x_1(t), \ldots, x_n(t)\big) , \end{cases} \tag{2.37}$$

or, shortly,

$$\begin{cases} \dot{x}_1 = f_1(x_1, \ldots, x_n) \\ \cdots = \cdots \\ \dot{x}_n = f_n(x_1, \ldots, x_n) . \end{cases} \tag{2.38}$$

By extension, we also call a *dynamical system* the differential system (2.35), or (2.37), or (2.38).

Existence and uniqueness of the integral curves are consequences of the Cauchy-Lipschitz Theorem [3].

Theorem 2.13 (Cauchy-Lipschitz) *For all vector fields and all points x_0 in the open set $\mathbb{X}$, there exists a unique maximal integral curve passing by x_0. We denote by $(\Phi_t(x_0), t \in I(x_0))$ this maximal integral curve, where $I(x_0)$ is an interval containing 0 and where $\Phi_0(x_0) = x_0$. Then, we have that*

1. *the set $\bigcup_{x_0 \in \mathbb{X}} I(x_0) \times \{x_0\}$ is an open set of $\mathbb{R} \times \mathbb{X}$;*
2. *the following mapping*

$$(t, x) \in \bigcup_{x_0 \in \mathbb{X}} I(x_0) \times \{x_0\} \mapsto \Phi_t(x) \in \mathbb{X} , \tag{2.39}$$

called local flow, *is smooth and satisfies*

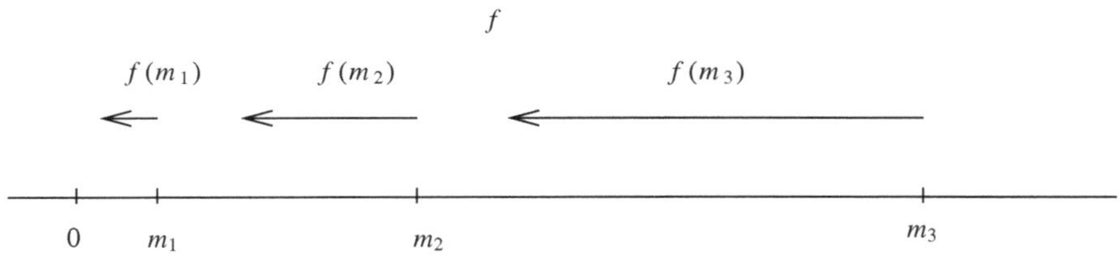

Fig. 2.9 Radioactive decay velocity $f(m)$ as a function of quantity of matter m

$$\frac{d}{dt}\Phi_t(x) = f\big(\Phi_t(x)\big) \; ; \tag{2.40}$$

3. *for all* $x_0 \in \mathbb{X}$ *and* t, s, $t+s$ *in* $I(x_0)$, *we have the* flow property*:*

$$\Phi_{t+s}(x_0) = \Phi_t\big(\Phi_s(x_0)\big) \; . \tag{2.41}$$

The flow is said to be *local* because it does not form a priori a family of mappings $(\Phi_t)_{t\in\mathbb{R}}$ from the open set $\mathbb{X}$ to $\mathbb{X}$. Indeed, the interval I on which the maximal integral curve passing by a point is defined depends of this point (the interval $I(x_0)$ may depend upon x_0).

Definition 2.14 *A vector field such that all maximal integral curves are parameterized by* $I = \mathbb{R}$ *is said to be* complete. *In that case, the local flow (2.39) is defined over* $\mathbb{R} \times \mathbb{X}$, *is denoted by* $(\Phi_t)_{t\in\mathbb{R}}$ *and is called a* global flow.

Let us now give some examples of vector fields and of integral curves.

Example 2.15 Physics provides examples of fields, such as the gravitation field or the electromagnetic field created by a magnet. If we sprinkle iron fillings near a magnet, we materialize the field by its field lines. The existence of these lines is an illustration of the Cauchy-Lipschitz Theorem: field lines are integral curves. △

Example 2.16 Radioactive decay.

The velocity $f(m)$ of radioactive decay of a radioactive body of mass m is assumed to be proportional to the quantity of matter m, that is, $f(m) = -km$. The phase space is here:

$$\mathbb{X} = \{m \in \mathbb{R} \mid m > 0\} \; . \tag{2.42}$$

The law of decay writes:

$$\dot{m} = -km, k > 0 \; . \tag{2.43}$$

The vector field f on the half-line $\mathbb{X}$ is directed towards the point 0 and the velocity vector of the flow is proportional to m as in Fig. 2.9. △

Definition 2.17 *Consider a vector field* f *on an open set* $\mathbb{X}$, *and the notations of Theorem 2.13. The* trajectory *of a point* x *of* $\mathbb{X}$ *is the mapping*

$$t \in I(x) \mapsto \Phi_t(x) \in \mathbb{X} . \tag{2.44}$$

The orbit *of a point x of $\mathbb{X}$ is the set $\{\Phi_t(x), t \in I(x)\}$.*

A trajectory is a *parameterized curve* of the phase space, whereas an orbit is a *subset* of the phase space.

Remark 2.18 Two orbits cannot cross because of uniqueness of integral curves. Indeed, if two orbits cross at a point, then they must coincide, because they are both associated with the *unique* integral curve passing by this point. We call *phase portrait* the partition of the phase space $\mathbb{X}$ induced by the orbits. ◇

A physical system is said to be in *stationary state*, or in *steady state*, if this state does not evolve during time. Mathematically, this is expressed by the following definition and proposition.

Definition 2.19 *We call* equilibrium *point (or* singular point*) of the vector field f on $\mathbb{X}$ any point $x_E \in \mathbb{X}$ such that:*

$$f(x_E) = 0 . \tag{2.45}$$

Proposition 2.20 *Consider the notations of Theorem 2.13. If the point $x_E \in \mathbb{X}$ is an equilibrium point of the vector field f, then $I(x_E) = \mathbb{R}$ and x_E is invariant by the flow Φ_t, namely $\Phi_t(x_E) = x_E$ for all $t \in \mathbb{R}$.*

Proof The trajectory $x(t) = x_E$ is such that

$$\frac{dx}{dt}(t) = 0 = f(x_E) = f\big(x(t)\big) .$$

By Theorem 2.13 and uniqueness of the trajectory initiating from the point x_E, we deduce that $x(t)$ coincides with $\Phi_t(x_E)$, that is, $\Phi_t(x_E) = x_E$. □

From a local perspective, equilibrium points play a particular role, in contrast to the other points, called *regular points* (states x such that $f(x) \neq 0$). Indeed, regular points satisfy the so-called *Straightening Theorem*, that we introduce without proof (see [10, 50]) and that we illustrate by Fig. 2.10.

Theorem 2.21 *If x_0 is a regular point, namely if $f(x_0) \neq 0$, there exists a* local diffeomorphism *(local smooth bijection) around the point x_0 that transforms the orbits of f into straight lines.*

In contrast to regular points, the singular points of a vector field f display a variety of local phase portraits. In a neighborhood of a singular point, the first nonzero term of the Taylor expansion of the vector field f is generally linear, and we discuss in Chap. 4 how this linear term may be sufficient for a local analysis of the original phase portrait. This is why we now turn to the study of *linear* dynamical systems.

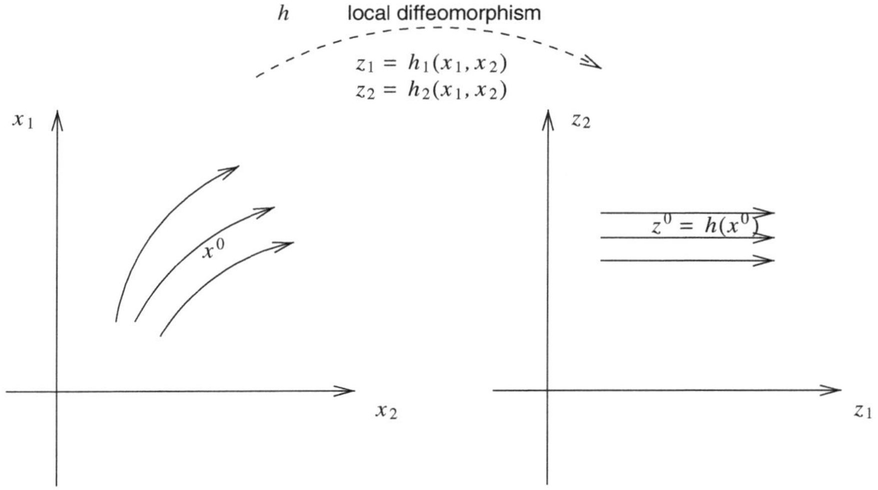

Fig. 2.10 Local straightening of the flow around a regular point

2.5 Linear Dynamical Systems

In what follows, A denotes a square matrix of size n *with real coefficients*. The matrix A defines a *linear dynamical system* on $\mathbb{R}^n$ by the vector field $f(x) = Ax$, and by the linear differential system:

$$\dot{x} = Ax \ . \tag{2.46}$$

The following proposition is classical [5, 18].

Proposition 2.22 *The matrix series*

$$e^A = \sum_{k=0}^{+\infty} \frac{1}{k!} A^k = I + A + \frac{1}{2!} A^2 + \frac{1}{3!} A^3 + \cdots \tag{2.47}$$

converges and defines a square matrix of size n. The vector field $x \in \mathbb{R}^n \mapsto Ax \in \mathbb{R}^n$ *is complete, and the flow of the linear differential system* (2.46) *is linear and is given by:*

$$\Phi_t(x) = e^{tA} x \ . \tag{2.48}$$

This result extends in any dimension the well known result of one dimensional linear differential equations, namely that

$$\dot{x} = \lambda x \iff x(t) = e^{\lambda t} x(0) \ . \tag{2.49}$$

However, if the behavior of the solution $x(t) = e^{\lambda t}x(0)$ is either exponential ($\lambda \neq 0$) or constant ($\lambda = 0$), the possible behaviors of the solution $\Phi_t(x) = e^{tA}x$ display more variety. The *real components* of the coefficients of the vector $e^{tA}x(0)$ include exponential, sinusoidal and polynomial expressions in the time variable t.

Example 2.23 The following computations of exponentials of matrices come either from computation of the series (2.47), or by direct resolution of the differential system (2.46) (see Exercises 2.6.5, 2.6.6 and 2.6.7):

$$A = \begin{pmatrix} \lambda_1 & 0 \\ 0 & \lambda_2 \end{pmatrix}, \quad e^{tA} = \begin{pmatrix} e^{\lambda_1 t} & 0 \\ 0 & e^{\lambda_2 t} \end{pmatrix}$$
$$A = \begin{pmatrix} 0 & 1 \\ -1 & 0 \end{pmatrix}, \quad e^{tA} = \begin{pmatrix} \cos t & \sin t \\ -\sin t & \cos t \end{pmatrix}$$
$$A = \begin{pmatrix} 0 & a \\ 0 & 0 \end{pmatrix}, \quad e^{tA} = \begin{pmatrix} 1 & at \\ 0 & 1 \end{pmatrix}.$$

△

By the expression (2.48), the flow of (2.46) is linear. Now, we highlight its structure, especially the existence of subspaces invariant by the flow, and we display analytical expressions. For this purpose, we start by recalling a result of linear algebra.

Definition 2.24 *If A is a square matrix of size n, we call* characteristic polynomial *the following polynomial of degree n*

$$\chi_A(s) = \det(sI - A) \ , \quad \forall s \in \mathbb{C} \ , \tag{2.50}$$

and $\mathcal{S}(A)$ the set of eigenvalues *of A, that is, the set of zeros of the polynomial χ_A, also called the* spectrum *of A. We call*

- multiplicity of an eigenvalue λ, *denoted by $m(\lambda)$, the multiplicity of the root λ of the characteristic polynomial χ_A;*
- index of an eigenvalue λ, *denoted by $\nu(\lambda)$, the first nonzero integer ν such that the following increasing sequence of subspaces $(\mathrm{Ker}(A - \lambda I)^\nu)_{\nu \geq 1}$ is stationary from $\nu = \nu(\lambda)$;*
- spectral subspace *associated with λ the complex kernel*

$$\mathcal{N}(\lambda) := \mathrm{Ker}(A - \lambda I)^{\nu(\lambda)} \subset \mathbb{C}^n, \tag{2.51}$$

that is,

$$x \in \mathcal{N}(\lambda) \iff \forall \nu \geq \nu(\lambda), (A - \lambda I)^\nu x = 0 \ .$$

Remark 2.25 If λ is an eigenvalue of A, one must take care to distinguish not only the index $\nu(\lambda)$ from the multiplicity $m(\lambda)$ (we have $1 \leq \nu(\lambda) \leq m(\lambda)$), but also the spectral subspace $\mathcal{N}(\lambda)$ from the *eigenspace* $\mathrm{Ker}(A - \lambda I) \subset \mathcal{N}(\lambda)$. ◇

The following proposition is classical [34].

Proposition 2.26 *If A is a square matrix of size n, then* $\mathbb{C}^n$ *can be decomposed into a direct complex sum of the spectral subspaces of A:*

$$\mathbb{C}^n = \bigoplus_{\lambda \in \mathcal{S}(A)} \mathcal{N}(\lambda) \,. \tag{2.52}$$

The projection p_λ *onto* $\mathcal{N}(\lambda)$ *is called the* spectral projection *of A at the eigenvalue* λ.

This result makes it possible to obtain an expression of the flow (2.48) projected on the *complex* decomposition (2.52) of the space $\mathbb{C}^n$.

Proposition 2.27 *For all* $x \in \mathbb{R}^n$, *we have that*

$$e^{tA}x = \sum_{\lambda \in \mathcal{S}(A)} \sum_{k=0}^{\nu(\lambda)-1} \frac{t^k}{k!} \exp(\Re(\lambda)t)(\cos \Im(\lambda)t + i \sin \Im(\lambda)t)(A - \lambda I)^k p_\lambda(x) \,. \tag{2.53}$$

Proof By linearity of the mapping $x \mapsto e^{tA}x$, it is sufficient to compute

$$\begin{aligned}
e^{tA} p_\lambda(x) &= e^{t(A-\lambda I)} e^{\lambda t} p_\lambda(x) \\
&= \sum_{k=0}^{+\infty} (A - \lambda I)^k \frac{t^k}{k!} e^{\lambda t} p_\lambda(x) \\
&= \sum_{k=0}^{\nu(\lambda)-1} (A - \lambda I)^k \frac{t^k}{k!} e^{\lambda t} p_\lambda(x) \text{ by the definition of } \nu(\lambda) \text{ in Definition 2.24} \\
&= \sum_{k=0}^{\nu(\lambda)-1} (A - \lambda I)^k \frac{t^k}{k!} e^{\Re(\lambda)t} (\cos \Im(\lambda)t + i \sin \Im(\lambda)t) p_\lambda(x) \,.
\end{aligned}$$

This ends the proof. □

We end with the following topological result [28] which will prove useful in subsequent chapters.

Proposition 2.28 *Let* $m_1, \ldots, m_k$ *be non zero integers summing to n* ($m_1 + \cdots + m_k = n$), *and let* $\mathcal{O}_1, \ldots, \mathcal{O}_k$ *be open subsets of* $\mathbb{C}$. *The set of matrices which have, taking into account their multiplicity,* m_1 *eigenvalues belonging to* $\mathcal{O}_1, \ldots, m_k$ *eigenvalues belonging to* $\mathcal{O}_k$, *is an open subset of the space of matrices.*

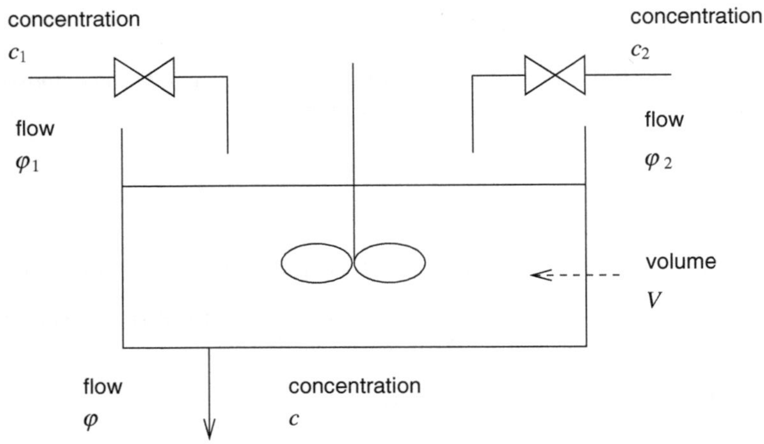

Fig. 2.11 Mixing process

2.6 Exercises

Exercise 2.6.1 Consider the chemical process consisting of mixing, in a vessel containing a solvent, two solutions of a same product having respective constant concentrations c_1 and c_2 and control flows φ_1 and φ_2 (see Fig. 2.11). We denote by V the volume of solution in the vessel, by c the concentration of output product and by g the output flow.

1. Under the assumption of incompressible fluids discussed in § 1.4.2, show that the balance equations of volume and of matter are of the form:

$$\begin{cases} \dfrac{d}{dt}V = \varphi_1 + \varphi_2 - g \\ \dfrac{d}{dt}(cV) = c_1\varphi_1 + c_2\varphi_2 - cg \,. \end{cases} \tag{2.54}$$

2. Suppose that $g = k\sqrt{\frac{V}{S}}$, where S is the constant section of the vessel and k is a parameter. Write the previous equations under the form of a state-model with state vector $x = (V, c)^\top$ and control vector $u = (\varphi_1, \varphi_2)^\top$.

Exercise 2.6.2 Consider the electrical circuit of Fig. 2.12.

1. By choosing as state variables the voltage x_1 at the terminals of the capacitor and the flow x_2 in the inductor, write a state-model of the system.
2. Write the temporal solution $x(t) = \big(x_1(t), x_2(t)\big)^\top$ of the state-model.

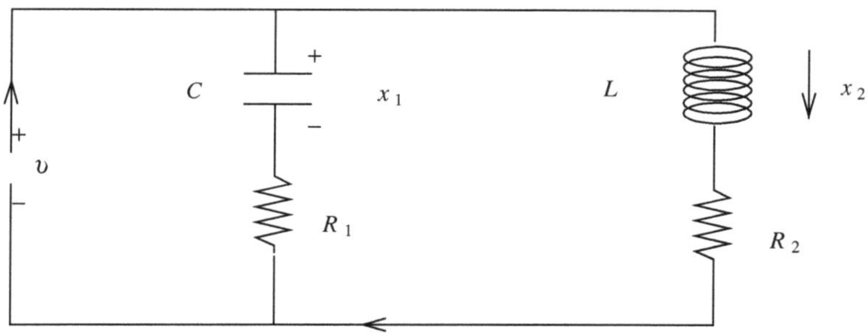

Fig. 2.12 Electrical circuit

Fig. 2.13 Ball on a rail

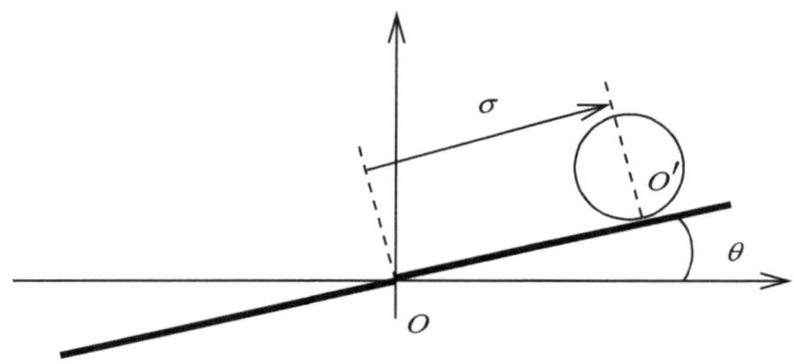

Exercise 2.6.3 We aim to regulate the position of a ball rolling on a rail, the inclination of which can be controlled by an motor (see Fig. 2.13).

We suppose that the ball rolls without slipping and we denote by:

- m the mass of the ball;
- r the radius of the ball;
- F the torque delivered by the motor;
- J_r the inertia of the rail;
- J_b the inertia of the ball.

From the Euler-Lagrange formalism discussed in § 1.4.1, determine the dynamical equations of the system rail + ball. Choose as state vector $x = (\sigma, \theta, \dot{\sigma}, \dot{\theta})^\top$.

Exercise 2.6.4 Show, by differentiating the expression (2.48) with respect to the time variable t, that (2.48) is indeed the flow of (2.46).

Exercise 2.6.5 If $A = \begin{pmatrix} \lambda_1 & 0 \\ 0 & \lambda_2 \end{pmatrix}$, compute e^{tA} by the formula (2.47).

Exercise 2.6.6 Let $A = \begin{pmatrix} 0 & 1 \\ -1 & 0 \end{pmatrix}$.

1. Check that $A^2 = -I$ and deduce the expression $e^{tA} = \begin{pmatrix} \cos t & \sin t \\ -\sin t & \cos t \end{pmatrix}$ by the formula (2.47).

2. Solve the differential system $\dot{x} = Ax$, that is,

$$\dot{x}_1 = x_2, \quad \dot{x}_2 = -x_1,$$

by expressing the solution as a function of the initial conditions $x_1(0)$ and $x_2(0)$. Deduce the expression of e^{tA}.

Exercise 2.6.7 Let $A = \begin{pmatrix} 0 & a \\ 0 & 0 \end{pmatrix}$.

1. Check that $A^2 = 0$ and deduce the expression $e^{tA} = \begin{pmatrix} 1 & at \\ 0 & 1 \end{pmatrix}$ by the formula (2.47).
2. Solve the differential system $\dot{x} = Ax$, that is,

$$\dot{x}_1 = ax_2 \ , \quad \dot{x}_2 = 0 \ ,$$

by expressing the solution as a function of the initial conditions $x_1(0)$ and $x_2(0)$. Deduce the expression of e^{tA}.

Chapter 3
Input-Output Representation

3.1 Introduction

For some systems, it may be difficult, or sometimes impossible, to obtain a mathematical dynamical model from physical laws. In that case, one can try to describe the dynamical behavior through so-called *input-output* relations, where the input signals correspond to perturbations or actions, and the outputs to system measurements. This "input-output representation" approach is also called the "frequency-domain" approach. It was intensively developed during the second World War, using the concept of a "black box" reacting to a set of input signals.

In § 3.2, we introduce input-output representations (a detailed mathematical treatment can be found in [63]). After general considerations, we turn the spotlight onto linear, causal and stationary representations, called *l.c.s. systems*. Because of this linearity, the role of the *Laplace transform* is emphasized (we refer the reader to § B.1 for recalls), and l.c.s. systems are characterized by *transfer matrices* whose elements are rational functions in the *Laplace variable* $s \in \mathbb{C}$. Rational functions being analytical, we know from Complex Analysis that they are uniquely determined by their values on the imaginary axis $i\mathbb{R} = \{i\omega \mid \omega \in \mathbb{R}\}$. The quantity $f = \omega/2\pi$ is interpreted as a *frequency* and the approach of working with functions of the Laplace variable s is called the *frequency-domain* approach. Due to their practical importance, the rest of the chapter is dedicated to the study of *scalar l.c.s. input-output representations*, also called *monovariable* and presented in § 3.3. Such scalar l.c.s. systems are characterized by *rational* transfer functions. A notion of *stability* is introduced in § 3.4, the so-called BIBO-stability, that is, *bounded input-bounded output*. It is discussed in terms of *poles* of the system, that is, the poles of the corresponding rational transfer function. Whereas the poles of a system are related to its stability, we show in § 3.5 that the *zeros* of a system are important with respect to disturbance rejection. The central ideas of *closed-loop* systems, of *controller* and of *feedback*, are introduced in § 3.6 under the specific form of the famous *PID compensator*, which stands for *proportional-integral-derivative*. Indeed, stabilization of a system is looked after by feeding the input with the output, thus "closing" the loop. The advantage of the

B. d'Andréa-Novel and M. De Lara, *Control Theory for Engineers*,
DOI: 10.1007/978-3-642-34324-7_3, © Springer-Verlag Berlin Heidelberg 2013

frequency-domain approach described is that it allows the construction of stabilizing compensators with natural definitions of degrees of robustness such as gain or phase margins through Nyquist and Bode diagrams. This is explained in § 3.7, and lead and lag phase compensators are discussed in § 3.8.

3.2 Input-Output Representation

We now give a definition of an input-output representation. A detailed mathematical treatment can be found in [63].

3.2.1 Definitions and Properties

Let m and p be two positive integers. Denote by $C^0(\mathbb{R}, \mathbb{R}^m)$ the space of continuous functions on the real line with values in $\mathbb{R}^m$, and by $l(\mathbb{Z}, \mathbb{R}^m)$ the space of sequences indexed by $\mathbb{Z}$ with values in $\mathbb{R}^m$. We call any element of $C^0(\mathbb{R}, \mathbb{R}^m)$ or $l(\mathbb{Z}, \mathbb{R}^m)$ an *input trajectory*, and any element of $C^0(\mathbb{R}, \mathbb{R}^p)$ or $l(\mathbb{Z}, \mathbb{R}^p)$ an *output trajectory*.

A *dynamical system* (Σ) with m input variables and p output measurements is a mapping from $C^0(\mathbb{R}, \mathbb{R}^m)$ towards $C^0(\mathbb{R}, \mathbb{R}^p)$ in continuous-time, and from $l(\mathbb{Z}, \mathbb{R}^m)$ towards $l(\mathbb{Z}, \mathbb{R}^p)$ in discrete-time. The input-output behavior is the relation

$$y = \Sigma(u). \tag{3.1}$$

Remark 3.1 In accordance with Remark 2.4, the notations u and y correspond here to *trajectories*, that is, elements of $C^0(\mathbb{R}, \mathbb{R}^m)$ and $C^0(\mathbb{R}, \mathbb{R}^p)$ in continuous-time, and $l(\mathbb{Z}, \mathbb{R}^m)$ and $l(\mathbb{Z}, \mathbb{R}^p)$ in discrete-time. However, this does not preclude u and y to denote *vectors* in the sequel, depending on the context. ◊

For any input or output trajectory x, we set

$$x[-\infty, t] := (x(\tau), \tau \leq t) \text{ and } x[-\infty, t[:= (x(\tau), \tau < t). \tag{3.2}$$

Dynamical systems deriving from physical laws are causal in the following sense.

Definition 3.2 *The dynamical system (Σ) is said to be* causal *(respectively,* strictly causal*), if the values of its output $y(t)$ at time t only depend on the values $u[-\infty, t]$ of its input u at times $\tau \leq t$ (respectively, $\tau < t$).*

A time translation of a trajectory $t \mapsto x(t)$ has the form $t \mapsto x(t - a)$.

Definition 3.3 *The dynamical system (Σ) is said to be* stationary *or* time-invariant *if its input-output behavior is invariant through time translation.*

Linear dynamical systems are of the upmost importance in Control Theory.

Definition 3.4 *The dynamical system* (Σ) *is said to be a* linear dynamical system *if its input-output behavior satisfies the following superposition principle: if the output vector* $y_i(t)$ *corresponds to the input vector* $u_i(t)$*, then each linear combination* $\sum_{i=1}^{l} \lambda_i u_i$ *of input vectors produces the output vector* $\sum_{i=1}^{l} \lambda_i y_i$.

In the rest of this chapter, we consider linear, causal and time-invariant (or stationary) dynamical systems.

Definition 3.5 *The dynamical system* (Σ) *is said to be* l.c.s. *if it is linear, causal and stationary.*

Remark 3.6 From now on, in the text, we only consider *continuous-time* l.c.s. system. Therefore, when not specified, l.c.s. system should be understood as continuous-time l.c.s. system. By causality and stationarity, all trajectories can and will be restricted to the domain $[0, +\infty[$. Appropriate assumption of continuity of the mapping (Σ) is also needed in what follows. ◊

3.2.2 Characteristic Responses and Transfer Matrices

Using the linearity property, we now show that every l.c.s. system (Σ) can be characterized by its response to particular classes of input signals.

Definition 3.7 *We call* Dirac delta function, Dirac distribution *or* unit impulse function *the distribution* δ *defined by*

$$\langle \delta, \varphi \rangle = \varphi(0) \ ,$$

for every smooth (C^∞) *test function* $\varphi : \mathbb{R} \to \mathbb{R}$ *with compact support, where* $\langle \cdot, \cdot \rangle$ *represents the duality product. We denote* δ_a *the distribution defined by* $\langle \delta_a, \varphi \rangle = \varphi(a)$.

Remark 3.8 The Dirac delta function is a mathematical object which makes it possible to describe a punctual density (of mass, or electrical...) or "distribution." In the engineering world, this impulse is generally introduced as follows:

$$\begin{cases} \delta(z) = 0 \text{ if } z \neq 0 \\ \delta(z) = +\infty \text{ if } z = 0 \\ \displaystyle\int_{-\infty}^{+\infty} \delta(z)\,\mathrm{d}z = 1 \ . \end{cases}$$

This is not mathematically correct if the previous integral is considered in the Lebesgue sense. However, the Dirac distribution is perfectly defined in the context of the so-called Theory of Distributions, elaborated by the French mathematician

Laurent Schwartz (see [60]). Distributions are *generalized functions*, which include the usual (locally integrable) functions. Under this theory, it is possible to introduce the "derivative of a distribution" and the Laplace or Fourier transform of a "tempered" distribution. Consequently, every distribution is differentiable and, therefore, every continuous function becomes differentiable. Moreover, the Fourier transform of any tempered distribution can be defined, which covers a much larger class than integrable or square-integrable functions for which the Fourier transform was originally defined. ◊

The Dirac delta function is the identity for the convolution operation (see § B.1 for the convolution definition):

$$u(t) = (u \star \delta)(t) = (\delta \star u)(t) = \int_{-\infty}^{+\infty} u(t-\tau)\delta(\tau)\,\mathrm{d}\tau\ .$$

Therefore, if $h = \Sigma(\delta)$ denotes the response of the time-invariant linear system (Σ) to the unit impulse function δ, the superposition principle (and an appropriate assumption of continuity of the mapping (Σ)) makes it possible to conclude that the response $y(t)$ to an arbitrary input trajectory u is given by:

$$y(t) = \int_{-\infty}^{+\infty} u(t-\tau)h(\tau)\,\mathrm{d}\tau = (h \star u)(t)\ .$$

For an l.c.s. system, causality makes it possible to write

$$y(t) = \int_{0}^{+\infty} u(t-\tau)h(\tau)\,\mathrm{d}\tau = (h \star u)(t)\ . \tag{3.3}$$

This motivates the following definition.

Definition 3.9 *The* impulse response *of an l.c.s. system* (Σ) *is the response* $h = \Sigma(\delta)$ *of this system to the unit impulse function* δ.

Remark 3.10 By (3.3),The impulse response is a condensed way to represent the dynamics of an l.c.s. system, since the system response to an arbitrary input u can be obtained through the convolution product of this input u with the impulse response h of the system. But impulse signals are difficult to realize. From a practical point of view, they can be numerically approximated by functions such as Gaussian functions $g_\sigma(z) = e^{-z^2/2\sigma^2}/\sigma\sqrt{2\pi}$ which tend towards δ (in the sense of distributions) when σ tends towards 0. ◊

Other classical signals are frequently used in Control Theory, such as the *unit step* and the *ramp* as defined below.

Definition 3.11 *We call*

- Heaviside step function *or* unit step function *the function* $\mathfrak{E}$ *defined by*

$$\mathfrak{E}(t) = 0 \ \ if \ \ t < 0 \ , \ \mathfrak{E}(t) = 1 \ \ if \ \ t \geq 0 \ , \tag{3.4}$$

- ramp *the function* $\mathfrak{R}$ *defined by*

$$\mathfrak{R}(t) = 0 \ \ if \ \ t < 0 \ , \ \mathfrak{R}(t) = t \ \ if \ \ t \geq 0 \ . \tag{3.5}$$

Definition 3.12 *The* step response *of an l.c.s. system, is the response of this system to the unit step function input* $\mathfrak{E}$.

Remark 3.13 In the context of distributions, the Dirac delta function δ is the derivative of the unit step input $\mathfrak{E}$, itself being the derivative of the ramp $\mathfrak{R}$ (see [60]). ◊

In general, even if the impulse response is known, the computation of outputs using convolution products is not easy. This is why *operational calculus* has been so widely used in the context of Control Theory: Laplace transform for linear continuous-time systems and z-transform for linear discrete-time systems. These transforms have the convenient property of transforming a convolution product into a simple product. The notion of transfer matrix is then introduced.

Definition 3.14 *The* transfer matrix *of a continuous-time l.c.s. system, with* m*-input vector* u *and* p*-output vector* y*, is the* $p \times m$ *matrix* $H(s)$ *such that*

$$Y(s) = H(s)U(s) \ , \tag{3.6}$$

where s *denotes the Laplace complex variable,* $U(s)$ *(respectively,* $Y(s)$*) the Laplace transform of* $u(t)$ *(if it exists), (respectively, of* $y(t)$*).*

Remark 3.15 In all rigour, we should speak of the transfer matrix *function* H (without the complex argument s), but we will currently speak of $H(s)$. The same holds true for $Y(s)$ and $U(s)$. ◊

Recall that a rational function is the ratio of two polynomials. It is said to be *proper* (respectively, *strictly proper*) if the degree of the numerator is less or equal than the degree of the denominator (respectively, strictly less than the degree of the denominator).

The proof of the following proposition is established in Chap. 5 in the case of continuous-time l.c.s. systems, and in Chap. 6 for discrete-time l.c.s. systems, using the state-space representation of linear, causal and time-invariant dynamical systems.

Proposition 3.16 *The elements of transfer matrices of l.c.s. systems are proper rational functions in the Laplace complex variable* s.

Let us point out that causality of an l.c.s. system is linked to the property of its associated transfer to be proper (cf. § 5.7 for continuous-time l.c.s. systems, § 6.9 for discrete-time l.c.s. systems and Remark 3.22 below).

Definition 3.17 *A transfer matrix is* proper *or* causal *if its elements are proper rational functions.*

From the definition of the impulse response and the fact that the Laplace transform of a convolution product is a simple product, the following proposition can be deduced.

Proposition 3.18 *The transfer matrix of an l.c.s. system is the Laplace transform of the impulse response.*

3.3 Single-Input Single-Output l.c.s. Systems

Due to their practical importance, the rest of the chapter is dedicated to the study of scalar systems, also called *monovariable* systems.

Definition 3.19 *A* scalar l.c.s. system *(or* monovariable system*) is an l.c.s. dynamical system with one input and one output, also called* single-input, single-output *system.*

Remark 3.20 In that case, the transfer matrix of Definition 3.14 is reduced to a *transfer function* which is a proper rational function. ◇

The following definitions are very useful in practice.

Definition 3.21 *A scalar l.c.s. system is said to be of* first-order *if its transfer function is of the form*

$$H(s) = \frac{k}{1 + Ts} \,. \tag{3.7}$$

A scalar l.c.s. system is said to be of second-order *if its transfer function is of the form*

$$H(s) = \frac{\omega_n^2}{s^2 + 2\zeta\omega_n s + \omega_n^2} \,, \tag{3.8}$$

where ζ *is called the* damping factor *and* $f_n = \omega_n/2\pi$ *the* natural frequency *of the system.*

For positive ζ and ω_n, the case $\zeta > 1$ corresponds to a so-called *hyperdamped* system (that is, with two real distinct and strictly negative roots), the case $\zeta = 1$ to a double real root equal to $-\omega_n$, and the case $\zeta < 1$ to two complex conjugated roots $-\zeta\omega_n \pm i\omega_n\sqrt{1-\zeta^2}$.

We now study the form of the impulse response of a scalar l.c.s. system in the time domain, with a proper transfer function (not necessarily strictly proper). Let $H(s)$ be such a rational transfer function (see Proposition 3.16):

$$H(s) = \frac{b_0 s^n + b_1 s^{n-1} + \cdots + b_n}{s^n + a_1 s^{n-1} + \cdots + a_n} .$$

After dividing the numerator by the denominator, we obtain:

$$H(s) = b_0 + \frac{\overline{b}_1 s^{n-1} + \cdots + \overline{b}_n}{s^n + a_1 s^{n-1} + \cdots + a_n} . \tag{3.9}$$

Thus, the transfer function $H(s)$ is the sum of a constant term b_0 and of a strictly proper transfer function $\overline{H}(s)$ which can be decomposed into a sum of simpler fractions as follows,

$$\overline{H}(s) = \overline{H}_1(s) + \cdots + \overline{H}_r(s) , \tag{3.10}$$

where

$$\overline{H}_i(s) = \frac{\alpha_1^i}{s - s_i} + \frac{\alpha_2^i}{(s - s_i)^2} + \cdots + \frac{\alpha_{\nu_i}^i}{(s - s_i)^{\nu_i}} . \tag{3.11}$$

The s_i are the roots of the denominator of $H(s)$, with multiplicity ν_i, such that $\sum_{i=1}^r \nu_i = n$. The Laplace transform being injective, it can be deduced from (3.9)–(3.11) and Table B.1 of Laplace transforms in § B.1—in particular, the relation $\mathcal{L}\left[e^{at}\frac{t^{m-1}}{(m-1)!}\mathfrak{E}(t)\right](s) = 1/(s-a)^m$—that the impulse response h of the system can be written as

$$h(t) = b_0\delta(t) + \overline{h}(t) \text{ with } \overline{h}(t) = \sum_{i=1}^{r}\left(\alpha_1^i + \alpha_2^i t + \cdots + \alpha_{\nu_i}^i \frac{t^{\nu_i - 1}}{(\nu_i - 1)!}\right)e^{s_i t} , \tag{3.12}$$

where we recall that $\delta(t)$ denotes the Dirac delta function introduced in Definition 3.7.

Remark 3.22 Let us point out that the constant term b_0 in the transfer function $H(s)$ given by (3.9) corresponds, in the time domain, to the term $b_0\delta(t)$ which leads to a direct relation between $u(t)$ and $y(t)$, since $y(t) = (h \star u)(t) = b_0 u(t) + (\overline{h} \star u)(t)$, $\delta(t)$ being the identity for the convolution operation. Therefore the system is strictly causal when $b_0 = 0$, that is when the transfer function $H(s)$ is strictly proper. ◇

Definition 3.23 *Consider the decomposition* (3.11) *of the Laplace transform $H(s)$ of the impulse response $h(t)$. If all the s_i have a strictly negative real part, we call* time constant *of the l.c.s. system with impulse response $h(t)$ given by* (3.12)*, the quantity σ defined by:*

$$\frac{1}{\sigma} = \min(\lambda_1, \ldots, \lambda_r) . \tag{3.13}$$

3.4 Stability and Poles: Routh's Criteria

Let us now define the notion of BIBO-stability, that is, *bounded input-bounded output-stability*.

Definition 3.24 *An l.c.s. system* (Σ) *is said to be* BIBO-stable *if, for all bounded input, the output remains bounded:*

$$\sup_{t\geq 0} \|u(t)\| < +\infty \Rightarrow \sup_{t\geq 0} \|\Sigma(u)(t)\| < +\infty\,. \tag{3.14}$$

Using similar arguments to those to be developed in § 5.3 and 6.4, the following proposition can be proven.

Proposition 3.25 *If the real parts of the characteristic roots of the transfer function of an l.c.s. system* (Σ) *are strictly negative, then* (Σ) *is* BIBO-*stable.*

We now give practical algebraic criteria that allows us to check if the roots of a polynomial have strictly negative real part (or a modulus strictly less than unity in discrete-time). The system stability can then be tested without computing the characteristic roots.

The Routh Criterion

Let $P(s)$ be a polynomial with real coefficients:

$$P(s) = a_0^1 s^n + a_0^2 s^{n-1} + a_1^1 s^{n-2} + a_1^2 s^{n-3} + a_2^1 s^{n-4} + a_2^2 s^{n-5} + \cdots\,. \tag{3.15}$$

The *Routh table* is constructed recursively

$$\begin{array}{llll} a_0^1 & a_1^1 & a_2^1 & \ldots \\ a_0^2 & a_1^2 & a_2^2 & \ldots \\ a_0^3 & a_1^3 & a_2^3 & \ldots \\ a_0^4 & a_1^4 & \ . & \ldots \\ . & . & . & \ldots \end{array} \tag{3.16}$$

using the formula

$$a_j^{i+2} = a_{j+1}^i - \frac{a_0^i}{a_0^{i+1}} a_{j+1}^{i+1}\,, \tag{3.17}$$

until no more term can be computed. The following theorem states the *Routh criterion* [58].

Theorem 3.26 *All the characteristic roots of the polynomial $P(s)$ in* (3.15) *have strictly negative real part if and only if all the coefficients a_0^i in the first column of the Routh table* (3.16) *given by* (3.17) *are nonzero and display the same sign. If all the a_0^i are nonzero, the number of coefficients in the first column of the Routh table having sign different from the sign of a_0^1 is the number of roots of the polynomial $P(s)$ with a strictly positive real part.*

A necessary condition for the polynomial $P(s)$ to have all its roots with a strictly negative real part is that all the coefficients a_i^j of $P(s)$ are nonzero and with the same sign. This condition is known as the *Hurwitz criterion* [40]. The Hurwitz criterion is only a necessary condition as shown by the following example.

Example 3.27 Consider the polynomial

$$P(s) = s^3 + 0.5s^2 + 3s + 3.5\ ,$$

which clearly satisfies the Hurwitz criterion. The corresponding Routh table is given by:

$$\begin{array}{cc} 1 & 3 \\ 0.5 & 3.5 \\ -4 & 0 \\ 3.5 & 0\ . \end{array}$$

From the Routh criterion, $P(s)$ has a real strictly positive root. △

3.5 Zeros of a Transfer Function

Whereas the poles of a system are related to its stability as discussed in § 3.4, we now discuss the importance of the "zeros" of a system with respect to disturbance rejection.

Definition 3.25 *The* zeros *of an l.c.s. system with transfer function $H(s)$ are the roots of the numerator of $H(s)$.*

Proposition 3.29 *If a* BIBO-*stable l.c.s. system* (Σ) *is subjected to a sinusoidal input of pulsation ω, the asymptotic output is a sinusoidal signal with the same pulsation ω, with amplification $|H(i\omega)|$ and with phase shift given by the angle $Arg\,(H(i\omega))$.*

Proof Let $h(t)$ be the impulse response of the l.c.s. system (Σ) and $H(s)$ be its transfer function. Suppose that the system is excited by a periodic input of pulsation ω of the form:

$$u(t) = e^{i\omega t} \text{ if } t \geq 0,\ u(t) = 0 \text{ else.} \tag{3.18}$$

By (3.3), the expression of the output $y(t)$ is given by

$$\begin{aligned}
y(t) &= \int_0^t h(\tau)e^{i\omega(t-\tau)}\,\mathrm{d}\tau \\
&= e^{i\omega t}\int_0^t h(\tau)e^{-i\omega\tau}\,\mathrm{d}\tau \\
&= e^{i\omega t}\int_0^{+\infty} h(\tau)e^{-i\omega\tau}\,\mathrm{d}\tau - e^{i\omega t}\int_t^{+\infty} h(\tau)e^{-i\omega\tau}\,\mathrm{d}\tau \;.
\end{aligned}$$

The system being BIBO-stable, the integral $\int_0^{+\infty} h(\tau)e^{-i\omega\tau}\,\mathrm{d}\tau$ exists and, by definition of the Laplace transform (B.1), we can write:

$$y(t) = e^{i\omega t}H(i\omega) - e^{i\omega t}\int_t^{+\infty} h(\tau)e^{-i\omega\tau}\,\mathrm{d}\tau \;.$$

The integral being convergent, $\int_t^{+\infty} h(\tau)e^{-i\omega\tau}\,\mathrm{d}\tau$ tends towards 0 when t tends towards infinity, which makes it possible to conclude. □

The following corollary is a straightforward consequence.

Corollary 3.30 *Consider a continuous-time l.c.s. system with transfer function $H(s)$. The zeros of $\omega \mapsto H(i\omega)$ correspond to the frequencies which are asymptotically rejected.*

Remark 3.31 The above ability to transform a periodic input signal into a periodic output signal with the same frequency, but amplified and dephased, is characteristic of linear systems. This is why the knowledge of the transfer function at various frequencies $f = \omega/2\pi$ is representative of the system. More precisely, the transfer function $H(s)$ being an analytical function, the knowledge of $H(s)$ on the imaginary axis—that is, of $H(i\omega)$ for all pulsations ω—is sufficient to determine $H(s)$ for all $s \in \mathbb{C}$. This is the reason why this approach is called the "frequency-domain analysis". ◊

Proposition 3.32 *Consider a continuous-time l.c.s. system with transfer function $H(s)$. If the transfer function $H(s)$ has a zero at $s = 0$, the step functions inputs of the form $b\mathfrak{E}$, where the Heaviside step function $\mathfrak{E}$ is defined in* (3.4)*, are asymptotically rejected.*

Proof Consider the continuous-time l.c.s. system with transfer function $H(s)$ excited by a step input signal with amplitude b. By Table B.1 of Laplace transforms in § B.1—in particular, the relation $\mathcal{L}[\mathfrak{E}(t)](s) = 1/s$—the Laplace transform of the output can be written as

$$Y(s) = H(s)\frac{b}{s} \;.$$

Applying the final value Theorem L7 in § B.1, we have that

$$\lim_{t\to+\infty} y(t) = \lim_{s\to 0} sY(s) = \lim_{s\to 0} bH(s) = bH(0) ,$$

which concludes the proof. □

This motivates the following definitions.

Definition 3.33 *We call* static gain *of the continuous-time l.c.s. system with transfer function* $H(s)$ *the real number* $H(0)$.

In § 3.7, we discuss why it is interesting to know if the zeros of a continuous-time l.c.s. system are in the left half complex plane (see also Exercise 3.9.5). Let us then introduce the following definition.

Definition 3.34 *A continuous-time l.c.s. system is said to be* minimum phase *if its zeros are all in the left half complex plane* $\{s \in \mathbb{C} \mid \Re(s) < 0\}$.

3.6 Controller Synthesis: The PID Compensator

Consider an l.c.s. system (Σ) which is not BIBO-stable. Our goal here is to transform the so-called *open-loop* system (Σ) by introducing a new auxiliary input and by "closing" the system with a so-called *feedback control law*, yielding a BIBO-stable *closed-loop* system.

From the input-output representation of an l.c.s. system discussed in § 3.2, two main types of controllers can be elaborated in the frequency-domain.

- The *precompensator* is represented on Fig. 3.1 and can be written, in the frequency-domain, as

$$U(s) = G(s)V(s) , \tag{3.19}$$

  where $V(s)$ represents a so-called *auxiliary input*. Therefore, the transfer function of the controlled system is of the form

$$Y(s) = H(s)G(s)V(s) . \tag{3.20}$$

- The *compensator* is represented on Fig. 3.2 and can be written as

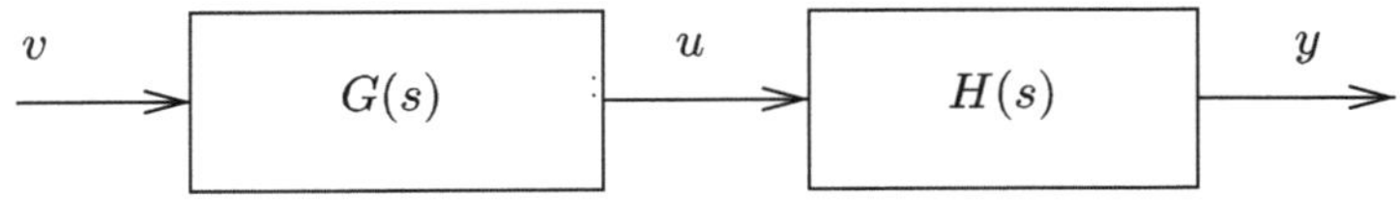

Fig. 3.1 The precompensator

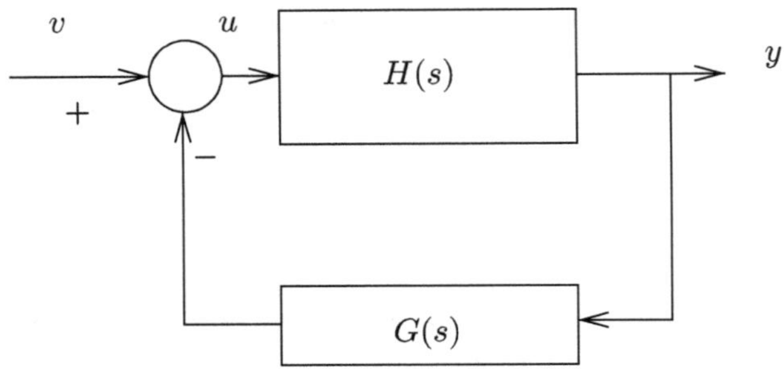

Fig. 3.2 The compensator

$$U(s) = -G(s)Y(s) + V(s)\ . \tag{3.21}$$

Combining (3.21) and $Y(s) = H(s)U(s)$, we obtain $Y(s) = H(s)V(s) - H(s)G(s)Y(s)$, and the transfer function of the closed-loop system is given by

$$Y(s) = \frac{H(s)}{1 + H(s)G(s)} V(s)\ . \tag{3.22}$$

The control laws (3.20) and (3.21) modify the transfer function of the system. If the open-loop system with transfer function $H(s)$ has unstable poles, using the precompensator technique requires that $G(s)$ is zero at the unstable poles of $H(s)$ to make the closed-loop system stable. This is not so easy since the poles of $H(s)$ are not known with perfect precision. On the other hand, the compensator technique is well adapted to stabilization, because it makes possible to place the closed-loop poles using $G(s)$, without necessarily knowing sharply the poles of $H(s)$.

Proposition 3.35 *Consider a continuous-time l.c.s. system with transfer function* $H(s)$. *To* BIBO-*stabilize the system with the compensator* (3.21), *it is necessary and sufficient to choose the coefficients of the compensator such that the characteristic roots of the polynomial* $1 + H(s)G(s)$ *have strictly negative real part.*

We now turn the spotlight onto the compensation technique called PID, for *proportional-integral-derivative*, which is still today much used in industry, on account of its simplicity.

Definition 3.36 *A* PID *control law is a compensator* (3.21) *with transfer function*

$$G(s) = K_P + K_D s + \frac{K_I}{s}\ , \tag{3.23}$$

where K_P *is the* proportional gain, K_D *the* derivative gain *and* K_I *the* integral gain.

3.6.1 First-Order Open-Loop System

Consider a first-order system as in (3.7) with an open-loop transfer function of the form:

$$H(s) = \frac{1}{s-\lambda} .$$

Then, the closed-loop transfer function (3.22) with PID compensator (3.23) is given by:

$$\frac{H(s)}{1+H(s)G(s)} = \frac{s}{s^2(1+K_D)+s(K_P-\lambda)+K_I} .$$

The single *proportional* feedback term ($K_D = K_I = 0$) makes it possible to place the closed-loop pole using the proportional gain K_P and therefore to improve the stability of the system or to stabilize it, if it was not originally BIBO-stable.

Moreover, the *integral* feedback term ($K_I \neq 0$) makes it possible to place a root $s = 0$ at the numerator of the closed-loop transfer function. As already mentioned in Proposition 3.32, the existence of a zero at $s = 0$ implies that constant step perturbations are asymptotically rejected for the closed-loop system.

The *derivative* feedback term K_D is useless in that case.

3.6.2 Open-Loop Second-Order System

Consider a second-order system as in (3.8) with an open-loop transfer function of the form:

$$H(s) = \frac{1}{s^2} .$$

The reader can easily check that, in the case of such a second-order system, the only proportional feedback term is not sufficient to stabilize the system, and that it is necessary to design a proportional-derivative feedback controller to ensure stability. An integral term would be useful, as for first-order systems, to asymptotically reject constant step perturbations.

We now turn the spotlight onto graphical representations that are useful to define robustness notions.

3.7 Graphical Methods: Gain and Phase Margins—Stability-Precision Dilemma

We now provide the reader with basic robustness notions, using graphical representations such as so-called *Nyquist* and *Bode diagrams*.

▷ For more details on the subject, we refer the reader to [47, 56, 61].

Except for some extensions to multivariable systems, presented for example in [47, 56], all the notions related to graphical representations have been established for continuous-time scalar l.c.s. systems, and this is the case which is tackled here.

Consider a scalar l.c.s. system with transfer function $H(s)$ on which the following proportional *output feedback* with gain K is applied

$$U(s) = -K\big(Y(s) - Y_c(s)\big) , \tag{3.24}$$

where $Y_c(s)$ is the reference output to be followed.

Remark 3.37 Equation (3.24) is a particular case of the compensator Eq. (3.21), where the compensator's transfer function is $G(s) = K$ and the auxiliary input is $V(s) = KY_c(s)$. ◊

The closed-loop transfer function $H_{\mathrm{CL}}(s)$ between the desired reference output Y_c and the original output Y is of the form:

$$H_{\mathrm{CL}}(s) = \frac{H(s)K}{1 + H(s)K} . \tag{3.25}$$

Recalling that $H(s)$ is a rational function, it can be written as:

$$H(s) = \frac{N(s)}{D(s)} . \tag{3.26}$$

Replacing $H(s)$ by this last expression in (3.25), we obtain:

$$H_{\mathrm{CL}}(s) = \frac{N(s)K}{D(s) + N(s)K} . \tag{3.27}$$

Let us now explain qualitatively why the gain K in (3.24) must be chosen with some constraints to guarantee stability. Actually, if the gain K tends towards zero in (3.27), the closed-loop poles of the transfer function $H_{\mathrm{CL}}(s)$ in (3.25) are given by the characteristic roots of $D(s)$ in (3.26), that is, by the poles of the closed-loop system. If the gain K tends towards $+\infty$ in (3.27), the poles of the closed-loop system are given by the roots of $N(s)$, that is by the zeros of the open-loop system. The interest of mimimum phase systems is then clear (see Definition 3.34).

However, since $H(s)$ corresponds to the transfer of a system which is generally strictly proper, the number n_p of open-loop poles is strictly greater than the number n_z of open-loop zeros. Therefore, the following question can be asked: when K tends towards $+\infty$, where are located the $d = n_p - n_z$ closed-loop poles which do not tend towards the n_z open-loop zeros? In fact, these poles tend towards infinity and they have an expression of the form[1]

[1] The symbol O corresponds to the *Big-O* notation: $f(z) = \mathrm{O}\big(g(z)\big)$ as $z \to z_0$ if and only if there exist a positive constant α and a neighborhood V of z_0 such that $|f(z)| \leq \alpha |g(z)|, \forall z \in V$.

$$|K|^{1/d} z_i + \mathrm{O}(1) , \tag{3.28}$$

where z_i are the roots of order d of $-C_N/C_D$ (respectively C_N/C_D) if $K > 0$ (respectively $K < 0$), where C_N (respectively C_D) is the coefficient of the term of highest degree of $N(s)$ (respectively $D(s)$) in (3.26). This result can be obtained from the graphical asymptotic locus of the roots (see for example [33]) or directly by calculus (see [21, Annexe B] or Exercise 3.9.7).

Then, when K tends towards infinity, from (3.28), the closed-loop poles of $H_{\mathrm{CL}}(s)$ in (3.25) which do not tend towards the open-loop zeros of $H(s)$ can be located in the right half complex plane and therefore will be unstable if d is greater than 2.

In practice, the gain K should not be chosen too high. Consequently, there exists a natural notion of "gain margin". To ensure the closed-loop stability, one can see from (3.25) that solutions in the right half complex plane of the equation $1 + H(s)K = 0$ should not exist.

Remark 3.38 From the above discussion, it is clear that, to preserve stability, the gains K must not be chosen too high. On the other hand, if K is too small, some difficulties with precision can occur, as illustrated by the following example.

Consider a system with transfer function $H(s)$ excited by a unit step signal $Y_c(s) = 1/s$. From the final value Theorem L7 in § B.1, the asymptotic error e_∞ between the output and the step reference input can be written as:

$$e_\infty = \lim_{s \to 0} s \left(\frac{1}{s} - \frac{H(s)K}{1 + H(s)K} \frac{1}{s} \right) = \frac{1}{1 + H(0)K} .$$

Therefore, to make the asymptotic error e_∞ decrease, or equivalently, to increase the precision of the closed-loop system, the term K should be chosen sufficiently high. Consequently, there is a dilemma concerning the choice of the gain depending on whether it is more important to increase stability or precision. This problem is usually known as the *stability-precision dilemma* and is also discussed in § 5.10. ◊

The Nyquist graphical method to analyze stability is based on the graphical representation of transfer functions in the complex plane.

Definition 3.39 *The* Nyquist diagram *of a transfer function $H(s)$ is the curve $\omega \mapsto H(i\omega)$ represented in the complex plane and graduated with increasing pulsations ω varying from* 0 *to* $+\infty$.

As discussed in Remark 3.31, the transfer function $H(s)$ is the result of a Laplace transform, and is therefore an analytical function. Then, the knowledge of $H(s)$ on the imaginary axis is sufficient to determine $H(s)$ everywhere (see the introductory discussion in § 3.1).

Example 3.40 In Fig. 3.3, a Nyquist diagram of a first-order system with transfer function $H(s) = 1/(1 + s)$ is plotted in function of the frequency $f = \omega/2\pi$.

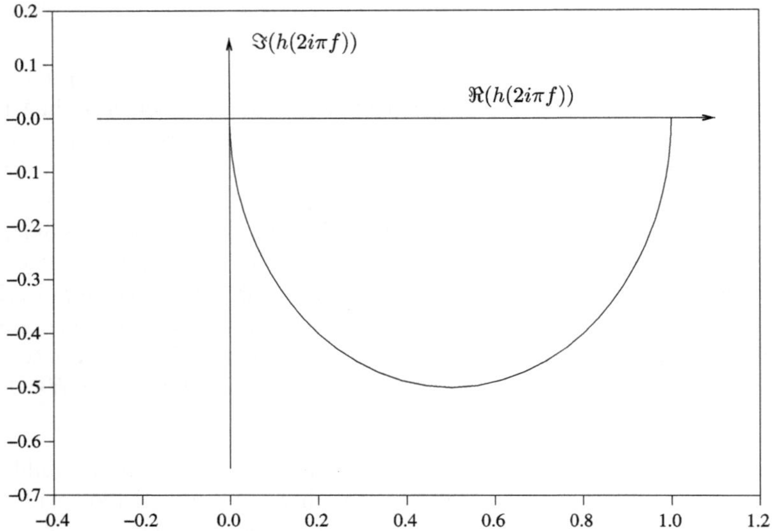

Fig. 3.3 Nyquist diagram of a first-order system

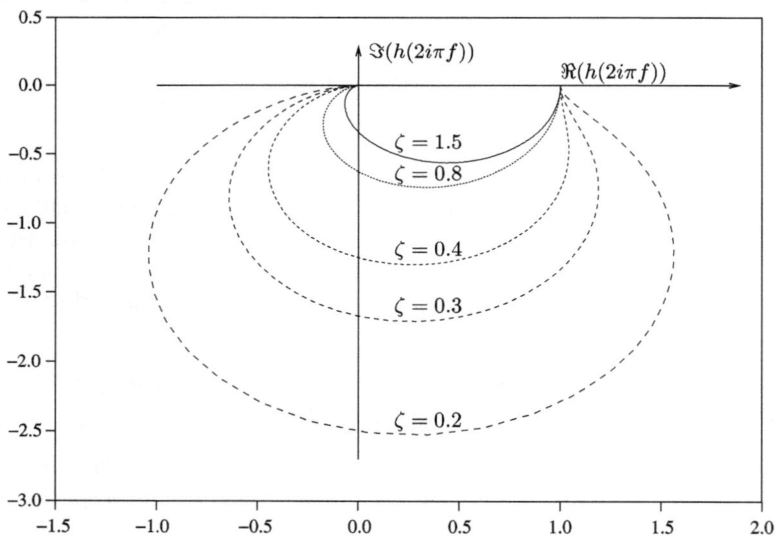

Fig. 3.4 Nyquist diagrams of second-order systems

In Fig. 3.4, Nyquist diagrams of second-order systems with transfer functions $H(s) = 1/(s^2 + 2\zeta s + 1)$ are plotted for different values of ζ, in function of the frequency $f = \omega/2\pi$.

Before giving the Nyquist stability result, let us recall a *Cauchy theorem* about curves in the plane (see [39, 42]).

Theorem 3.41 (Cauchy) *Consider $\mathcal{C}$ a simple closed complex curve, oriented in the clockwise direction and $H(s)$ a rational function with no zero and no poles on $\mathcal{C}$. Then, the following equality is satisfied*

$$N = P - Z \,, \tag{3.29}$$

where, if $H(\mathcal{C})$ denotes the image by $H(s)$ of the curve $\mathcal{C}$,

- *N is the number of encirclements (taken in the counterclockwise direction) around the origin $H(\mathcal{C})$;*
- *P is the number of poles of $H(s)$ counted with their multiplicity, which are inside $\mathcal{C}$;*
- *Z is the number of zeros of $H(s)$ counted with their multiplicity, which are inside $\mathcal{C}$.*

Consider a scalar l.c.s. system with transfer function $H(s)$, on which we apply a proportional feedback with unit gain, that is, $K = 1$. Using (3.25), the closed-loop poles are nothing but the zeros of $1 + H(s)$. To test the system stability, we now consider the closed complex curve $\mathcal{B}$, called *Bromwich contour*, represented on Fig. 3.5, where R tends towards $+\infty$.

Definition 3.42 *Consider a continuous-time l.c.s. system with transfer function $H(s)$. When R tends towards $+\infty$, the image by $H(s)$ of the Bromwich contour $\mathcal{B}$ is called the* Nyquist locus *of the l.c.s. system with transfer function $H(s)$.*

Then, from Cauchy Theorem 3.41, the stability of the closed-loop system depends on the number Z of zeros of $1 + H(s)$ which are located inside the Bromwich contour $\mathcal{B}$. Applying Cauchy Theorem 3.41 easily makes it possible to deduce the following stability result, known as the *Nyquist criterion*.

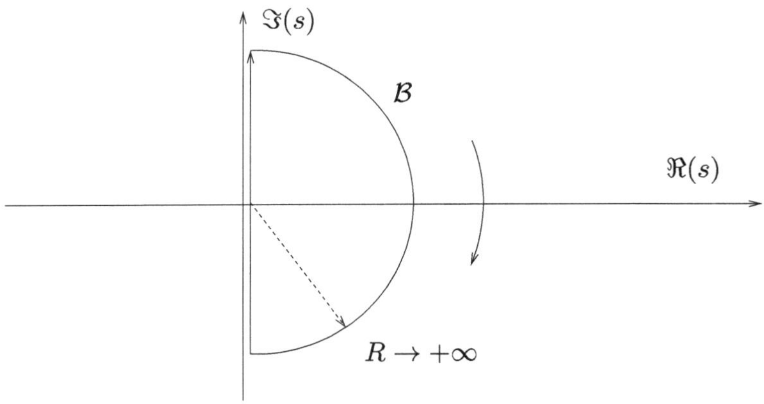

Fig. 3.5 Bromwich contour

Theorem 3.43 *The l.c.s. system with open-loop transfer function $H(s)$ is* BIBO*-stable in closed-loop with a unit proportional feedback ($K = 1$) if and only if its Nyquist locus encircles the point -1 in the counterclockwise direction as many times as the number of unstable open-loop poles.*

Remark 3.44 To build the Nyquist locus from the Nyquist diagram, the last one has to be completed symmetrically from the real axis to obtain the image by $H(s)$ of the strictly negative part of the imaginary axis $i\mathbb{R}^-$, and we have to plot the locus part corresponding to the image by $H(s)$ of the semi-circle of radius R, where R tends towards infinity (see Fig. 3.5). Generally, the systems we are considering have proper rational transfer functions, so that this last part of the image is reduced to a point in the complex plane. On the other hand, if $H(s)$ has poles at $s = 0$ or on the imaginary axis, another closed complex curve which avoids singularities has to be considered, in place of the Bromwich contour, to apply Cauchy Theorem 3.41 (see Exercise 3.9.6 or [33, Chap. 6]). ◊

If we had considered a general positive scalar gain K, the closed-loop transfer would have been given by (3.25). Multiplying by K is equivalent to applying to the Nyquist locus an homothety with center the origin and with ratio K. Therefore, if KA is strictly greater than 1, A being the gain corresponding to a dephasing of 180 degrees, the system becomes unstable as shown by Fig. 3.6.

Following the above discussion, we introduce the following definition.

Definition 3.45 *The* gain margin *of the l.c.s. system with transfer function $H(s)$ is the quantity $G_m = 1/A$, where A is the gain corresponding to a dephasing of 180 degrees.*

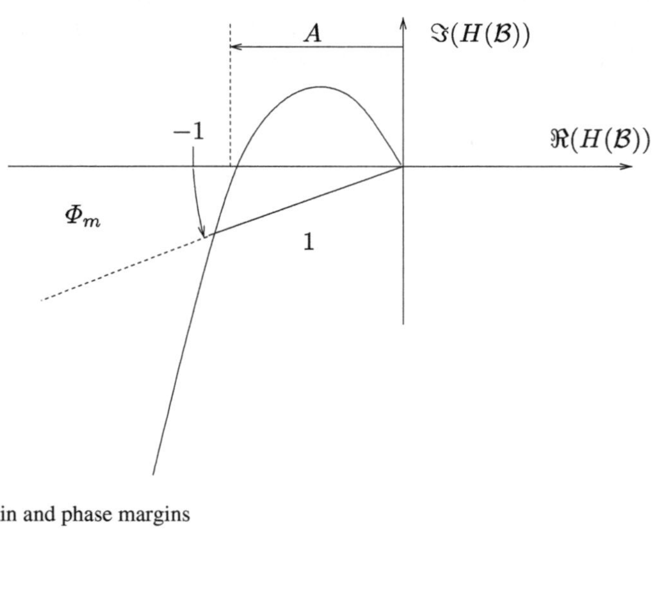

Fig. 3.6 Gain and phase margins

Similarly, the system can become unstable if the phase increases. Increasing the phase of an angle α is equivalent to rotate the Nyquist locus with the same angle α around the origin. Then, the Nyquist locus can reach the critical point -1 if it rotates with an angle Φ_m, corresponding to the angle between the horizontal axis and the line connecting the origin and the intersection point of the Nyquist locus and the circle of center the origin and radius 1 (see Fig. 3.6).

Definition 3.46 *The* phase margin *of the l.c.s. system with transfer function $H(s)$ is the angle Φ_m defined on Fig. 3.6.*

These gain and phase margins constitute a measure of the stability and *robustness* degree of the control with respect to disturbances on the transfer function, due for example to modelling errors or neglected dynamics.

The gain and phase margins can also be obtained from *Bode diagrams*, and we refer the reader to Appendix D for details concerning the construction of these diagrams.

Remark 3.47 Let us point out that a proportional output feedback is not sufficient to modify the phase of a system. More general PID compensators such as lead and lag compensators (presented in the next § 3.8) should also be considered. ◊

3.8 Lead and Lag Phase Compensators

Consider a compensator as in (3.21). If the transfer function $G(s)$ of the compensator is such that $G(i\omega)$ has a positive phase in the vicinity of the system's natural frequency (and associated pulsation) with transfer function $H(s)$, or in other words the pulsation for which $H(i\omega)$ is the closest to the critical point -1, such a compensator clearly improves stability and takes the following form:

$$G(s) = K\frac{s+a}{s+b}, \quad b > a > 0. \tag{3.30}$$

The transfer function $G(s)$ can also be rewritten as

$$G(s) = K\frac{s+1/T}{s+1/(\alpha T)} \text{ with } a = \frac{1}{T}, \quad b = \frac{1}{\alpha T}, \quad 0 < \alpha < 1.$$

The Bode diagram of $G(s)$ is symmetrical with respect to the geometric mean pulsation $\sqrt{ab}$, where the *lead phase* is precisely maximum. The compensator can then be tuned such that the maximum lead phase is located at the natural or resonance frequency of the system. The gain margin makes it possible to improve precision using the gain K while preserving stability (see Fig. 3.7).

Remark 3.48 Notice that, in the case where a takes small values close to 0 and b takes high values in (3.30), we obtain a "generalized derivator." ◊

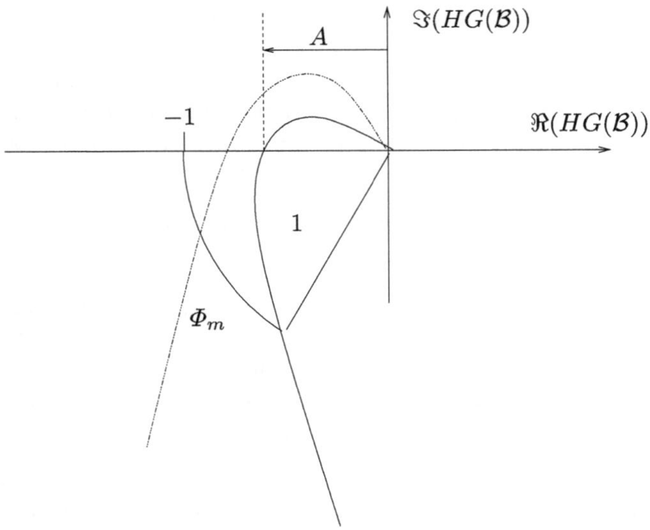

Fig. 3.7 Lead phase

In the same way, a *lag phase* compensator can be defined with a transfer function similar to (3.30), but such that $0 < b < a$.

In that case, if $G(s)$ is high for $s = 0$ (that means a high and b small) and if it introduces a negligible dephasing in the vicinity of the natural frequency of the system, the precision of the closed-loop system can be improved while preserving stability. It can then be interpreted as a "generalized integrator."

The industrial success of output feedback control laws, and particularly of PID compensators, is mainly due to their simplicity, because it is not critical to know sharply a state-space model of the system to tune the compensator's gains. Moreover, some multivariable systems can be decomposed in networks of first-order and second-order systems which can therefore be stabilized using local PID controllers. Nevertheless, this approach has drawbacks: the multiplication of local controllers can obscure the problem and be totally inefficient in the case of highly coupled physical phenomena, possibly leading to instability of the closed-loop system. This is the reason why global stabilization methods using state-space approach and described in the next chapters prove useful and efficient.

Moreover, as already mentioned in § 3.5, the zeros of a transfer function play an important role with respect to disturbance rejection and the input-output representation is better adapted to the computation of zeros. In Chap. 8, the notion of zeros is extended to multivariable systems which can be described by a transfer matrix, and some results from the theory of polynomial matrices are presented.

The poles being fixed, the interest of "astutely placing zeros" of a multivariable system by using remaining degrees of freedom in the control is pointed out with respect to disturbance rejection.

3.9 Exercises

Exercise 3.9.1 We consider and study transfer functions of electrical circuits.

1. Compute the transfer function of an electrical circuit with input voltage v and output charge q described in § 2.3.5. Then, do the same if the considered output is the electrical current i.
2. Compute the transfer function of an electrical circuit with input voltage v and output charge q described in Exercise 2.6.2.

Exercise 3.9.2 Consider the following transfer function:

$$H(s) = \frac{5(s+3)((s-2)^2+9)}{(s+4)((s+2)^2+4)((s+6)^2+1)} .$$

1. Determine the poles and zeros of $H(s)$ and draw them in the complex plane.
2. Determine the pole or the pair of complex conjugate poles corresponding to the time constant of the system (see Definition 3.23). How are they located with respect to the imaginary axis?
3. Give the form of the impulse response.
4. Plot the asymptotic Bode diagrams (see Appendix D).

Exercise 3.9.3 A continuous-time l.c.s. dynamical system with impulse response $h(t)$ is excited by a periodic input $u(t) = e^{i\omega t}\mathfrak{E}(t)$, where we recall that $\mathfrak{E}(t)$ denotes the Heaviside step function (3.4). What is the asymptotic behavior of the output?

Exercise 3.9.4 Plot the Nyquist and asymptotic Bode diagrams (see Appendix D) of a first-order l.c.s. system (3.7) for different values of T and of a second-order system (3.8) for different values of ζ and ω_n.

Exercise 3.9.5 Consider an l.c.s. system with the transfer function

$$H(s) = \frac{s-s_0}{(s+\lambda_1)(s+\lambda_2)} , \quad \lambda_1 > 0 , \quad \lambda_2 > 0 .$$

Compute the step response of this system to a step input with amplitude α. Show that it tends towards a step signal with sign the one of $-s_0\alpha$. Deduce that, if the system is minimum phase (see Definition 3.34), its step response tends towards a step signal having the same sign than the reference one.

Exercise 3.9.6 Applying the Nyquist criterion (see Theorem 3.43), test the closed-loop stability, by a unit proportional feedback, of the l.c.s. system with transfer function

$$H(s) = \frac{1+s}{s(1-s)} .$$

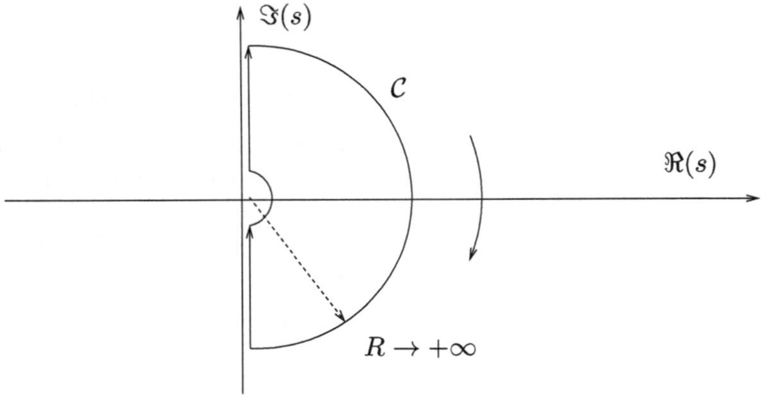

Fig. 3.8 Stability curve $\mathcal{C}$

For this purpose, the reader can consider the curve $\mathcal{C}$ of Fig. 3.8 (see the difference with Fig. 3.5).

Exercise 3.9.7 Consider the rational transfer function $H(s)$ of an l.c.s. system, given under the form (3.26). The problem considered here is the asymptotic behavior of the roots of the polynomial $D(s) + KN(s)$ when the gain K in the output feedback (3.24) tends towards infinity. This problem is equivalent to the asymptotic study of the roots of $N(s) + \varepsilon D(s)$ when $\varepsilon = 1/K$ tends towards zero. In what follows, n_p denotes the degree of $D(s)$, n_z the degree of $N(s)$ and $d = n_p - n_z$.

1. Consider a function $\mathfrak{F} : \mathbb{C} \times \mathbb{R} \to \mathbb{C}$ of the form

$$\mathfrak{F}(z, \eta) = \mathfrak{F}_1(z) + \eta \mathfrak{F}_2(z, \eta) ,$$

where $\mathfrak{F}_1(z)$, $\mathfrak{F}_2(z, \eta)$ are analytical functions of z and $\mathfrak{F}_2$ is assumed to be C^∞ in (z, η). Applying the implicit function theorem [10], first show that, if z_0 is a simple root of $\mathfrak{F}_1(z)$, then a C^∞ function ϕ exists such that, in a neighborhood of $(z_0, 0)$, we have that

$$\mathfrak{F}(z, \eta) = 0 \iff z = z_0 + \eta\phi(\eta) .$$

2. Case $d = 0$. Show that if ε tends towards 0, $N(s) + \varepsilon D(s)$ has n_z ($=n_p$) bounded roots which tend towards the roots of $N(s)$. To do this, item 1 and the Rouché theorem can be applied (see [39, 42]. The case where $N(s)$ has a root s_0 with multiplicity $\mu \geq 2$ can be considered by introducing the change of variable $s = s_0 + \varepsilon^{1/\mu} z$ to avoid the singularity.
3. Case $d > 0$. Show that if ε tends towards 0 with $\varepsilon > 0$, the d roots of $N(s) + \varepsilon D(s)$ which do not tend towards the n_z roots of $N(s)$ are of the form $\varepsilon^{-1/d} s_i + \mathrm{O}(1)$,

$i = 1, \ldots, d$. The s_i are the roots of order d of $-C_N/C_D$ where C_N (respectively, C_D) is the highest term's coefficient of $N(s)$ (respectively, $D(s)$).
The change of variable $s = \varepsilon^{-1/d}z$ can be considered and it can be shown that $N(s) + \varepsilon D(s) = 0$ can be rewritten as

$$N(s) + \varepsilon D(s) = \mathfrak{F}(z, \varepsilon^{1/d}) = C_N + C_D z^d + \varepsilon^{1/d} R(z, \varepsilon^{1/d}) = 0 \ ,$$

where $R(z, \varepsilon^{1/d})$ is analytical and is an O(1) when ε tends towards 0.
Finally, the result of item 1 can be applied to $\mathfrak{F}(z, \varepsilon^{1/d})$.

4. Now, applying the result of item 3, show that, if ε tends towards 0 with $\varepsilon < 0$, the roots of $N(s) + \varepsilon D(s)$ not tending towards the n_z roots of $N(s)$ are of the form $(-\varepsilon)^{-1/d} s_i + \mathrm{O}(1)$, $i = 1, \ldots, d$. The s_i are the roots of order d of C_N/C_D, which proves Eq. (3.28).

Part II
Stabilization by State-Space Approach

Chapter 4
Stability of an Equilibrium Point

4.1 Introduction

Stabilizing a system in a neighborhood of a steady state is one of the first goals of Control Theory. For this purpose, we cast a glow in § 4.2 on the notions of *stability* and of *asymptotic stability* of an equilibrium point for general dynamical systems as discussed in § 2.4. The case of linear dynamical systems is treated in § 4.3. For linear systems in the plane, we provide a detailed classification of the stability of the zero equilibrium in § 4.4. Then, we turn back to general nonlinear dynamical systems, and observe that, in an appropriate neighborhood of an equilibrium point, a vector field coincides with the linear term of its Taylor development up to the first-order. This is why, in § 4.5, we introduce the *tangent linear system* and we discuss to what extent the stability of an equilibrium point can be deduced from the stability properties of the tangent linear system. Stability theory in the sense of Lyapunov and, especially, *Lyapunov functions* are discussed in § 4.6. They provide stability criteria directly depending on the vector field, and not on the flow as in the general definitions of stability. Lyapunov functions make it also possible to achieve more global stability. At last, we consider *controlled* nonlinear dynamical systems, namely state-models as defined in § 2.2, and discuss how to stabilize them locally around an equilibrium point in § 4.7.

4.2 Stability and Asymptotic Stability of an Equilibrium Point

The equilibrium points of a dynamical system can be classified according to their "stability." Let us illustrate this concept using the pendulum example described in § 2.2. The state vector $x = (\theta, \dot{\theta})^\top$ has been introduced and the system's dynamics are given by the vector field

B. d'Andréa-Novel and M. De Lara, *Control Theory for Engineers*,
DOI: 10.1007/978-3-642-34324-7_4, © Springer-Verlag Berlin Heidelberg 2013

$$f(x) = f\begin{pmatrix}\theta \\ \dot{\theta}\end{pmatrix} = \begin{pmatrix}\dot{\theta} \\ -\frac{g}{l}\sin\theta\end{pmatrix}. \tag{4.1}$$

Intuitively, the pendulum's behavior is different depending on whether the system is close to the upper equilibrium position $x_{sup} = (\pi, 0)^\top$ or close to the lower one $x_{inf} = (0, 0)^\top$. In fact, near the upper x_{sup}, the gravitation tends to make the pendulum fall and oscillate indefinitely around the lower equilibrium position x_{inf}.

The equilibrium x_{sup} is said to be "unstable" because the trajectories tend to move away from x_{sup}, whereas x_{inf} is said to be "stable" because the trajectories remain in a neighborhood of x_{inf}. If we take into account friction effects due, for example, to air resistance, the pendulum tends to stop at the position x_{inf} which can then be called an "asymptotically stable equilibrium."

This example points out the fundamental role of equilibrium points regarding the evolution of dynamical systems. This evolution differs depending on whether the trajectories tend to move away from or closer to the equilibrium point which is considered. This fact characterizes the stability or instability property of an equilibrium point and these notions are now elaborated on (see [57]).

Consider the dynamical system (2.35). Equilibrium points are defined in (2.45), and the flow Φ_t in (2.39) in Theorem 2.13.

Definition 4.1 *An equilibrium point x_E of the dynamical system* (2.35) *is said to be*

- stable *if, for all neighborhood W' of x_E, a neighborhood W'' of x_E exists such that, for all x in W'', $\Phi_t(x)$ exists and belongs to W' for all $t \geq 0$ (an equilibrium is said to be* unstable *if it is not stable);*
- attractive *if a neighborhood W' of x_E exists such that, for all x in W', $\Phi_t(x)$ exists for all $t \geq 0$ and $\Phi_t(x)$ tends to x_E when $t \to +\infty$;*
- asymptotically stable *if it is both stable and attractive.*

To be stable means that the trajectory remains as close as expected to the equilibrium point x_E, as soon as the initial state is close enough to x_E. To be attractive means that the state trajectory converges to the equilibrium point x_E (see Fig. 4.1). Notice that stability is both a *transient* and *asymptotical* property, that considers the trajectory $t \mapsto \Phi_t(x)$ for all times $t \geq 0$, whereas attractivity is only an *asymptotical* property.

Definition 4.2 *The* basin of attraction *of an equilibrium point x_E of the dynamical system* (2.35) is the set of points x in the phase space $\mathbb{X}$ such that $\Phi_t(x)$ tends to x_E when $t \to +\infty$.

Remark 4.3 If the basin of attraction of an asymptotically stable equilibrium point x_E is equal to the whole phase space $\mathbb{X}$, the equilibrium point is said to be *globally asymptotically stable.* ◇

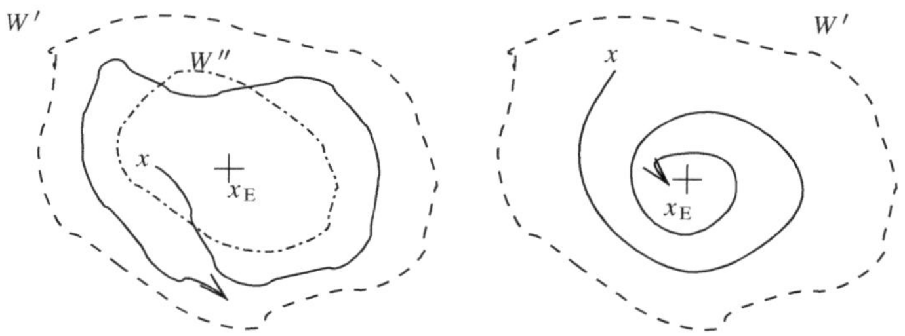

Fig. 4.1 Stable equilibrium and attractive equilibrium

4.3 The Case of Linear Dynamical Systems

For linear dynamical systems, the zero (origin) is always an equilibrium. We now highlight how the trajectories of the linear dynamical system (2.46) depend on the matrix A, and show that the asymptotic stability of the origin only depends on the *sign of the real part of the eigenvalues* of the matrix A. To do this, we use results of § 2.5, and the notations of Definition 2.24 and of Propositions 2.26 and 2.27.

Proposition 4.4 *Let A be a real square matrix of dimension n, with distinct eigenvalues $\lambda_1, \ldots, \lambda_r$. Let us consider the equilibrium point $x_E = 0$ of the linear dynamical system* (2.46).

1. *If there exists at least one $i = 1, \ldots, r$ such that $\Re(\lambda_i) > 0$, the equilibrium point 0 is unstable.*
2. *If for all $j = 1, \ldots, r$, one has that $\Re(\lambda_j) \leq 0$, then*
   (a) *if $\forall j = 1, \ldots, r$, $\Re(\lambda_j) < 0$, then the equilibrium point 0 is asymptotically stable;*
   (b) *if $\exists i = 1, \ldots, r$, $\Re(\lambda_i) = 0$ and $\nu(\lambda_i) > 1$, the equilibrium point 0 is unstable;*
   (c) *if $\exists i = 1, \ldots, r$, $\Re(\lambda_i) = 0$ and if $\forall j = 1, \ldots, r$, $(\Re(\lambda_j) = 0 \Rightarrow \nu(\lambda_j) = 1)$, the equilibrium point 0 is stable but not asymptotically stable.*

Proof In each case, the issue is one of comparing an exponential term $\exp(\Re(\lambda)t)$ to a polynomial term t^k in the expression (2.53).

1. Since $\text{Ker}(A - \lambda_i I)^{\nu(\lambda_i)} \backslash \text{Ker}(A - \lambda_i I)^{\nu(\lambda_i)-1}$ is not reduced to the singleton $\{0\}$ by definition of the eigenvalue index $\nu(\lambda_i)$, we choose a nonzero complex vector $z \in \mathcal{N}(\lambda_i) \backslash \text{Ker}(A - \lambda_i I)^{\nu(\lambda_i)-1}$. From z, we build a nonzero *real* vector $x = z + \bar{z}$ by using the conjugate mapping recalled in § 2.1. In (2.53) , we have that $p_\lambda(x) = 0$ when $\lambda \neq \lambda_i$, so that $\|e^{tA}x\|$ exponentially increases to $+\infty$, hence the equilibrium point 0 is unstable.
2. (a) Let μ be such that $\max\{\Re(\lambda_1), \ldots, \Re(\lambda_r)\} < \mu < 0$. We deduce from (2.53) that there exists a constant M such that

$$\|e^{tA}x\| \leq Me^{\mu t}\|x\|, \quad \forall x \in \mathbb{R}^n. \tag{4.2}$$

We conclude that the equilibrium point 0 is asymptotically stable.

(b) The vector x is chosen as in item 1. It can then be observed in (2.53) that $\|e^{tA}x\|$ grows as a polynomial $t^{\nu(\lambda_j)-1}$, where $\nu(\lambda_i)-1 \geq 1$, towards $+\infty$: hence the equilibrium point 0 is unstable.

(c) We observe in (2.53) that there exists a constant M such that $\|e^{tA}x\| \leq M\|x\|$, for all $x \in \mathbb{R}^n$: the equilibrium point 0 is thus stable. Then, we choose x as in item 1 and observe that $e^{tA}x$ does no tend to zero when t goes to $+\infty$. Therefore, the equilibrium point 0 is not asymptotically stable.

This ends the proof. □

Let us point out that stability, an issue of mathematical analysis—namely, the asymptotic behavior of the flow of a linear differential equation in Definition 4.1—has now been turned into an algebraic problem, the evaluation of eigenvalues of a matrix, thanks to Proposition 4.4. Thus, we are now equipped with a simple criterion to check the asymptotic stability of the equilibrium point $x_E = 0$ of the linear system (2.46), as follows.

Theorem 4.5 *The equilibrium point $x_E = 0$ of the linear system* (2.46) *is asymptotically stable if and only if the eigenvalues of the matrix A have strictly negative real part.*

Definition 4.6 *We call* stability half-plane *the set $\{s \in \mathbb{C} \mid \Re(s) < 0\}$ of complex numbers with strictly negative real part. A square matrix is said to be* asymptotically stable *if its eigenvalues belong to the stability half-plane.*

Remark 4.7 We also define a *stable matrix* and an *unstable matrix* in the corresponding cases of Proposition 2.28. ◇

The following topological result is a corollary of Proposition 2.28.

Proposition 4.8 *If a matrix is asymptotically stable, there exists a neighborhood in which all the matrices are asymptotically stable.*

In particular, if $(A(p))_{p\in P}$ is a family of square matrices continuously depending on a parameter p and if $A(p_0)$ is asymptotically stable, it is the same for $A(p)$ where p belongs to an appropriate neighbourhood of p_0.

In Control Theory, one traditionally calls *proper modes* or *poles* (see Definition 5.57) the eigenvalues of the square matrix attached to a linear system.

Definition 4.9 *We call* proper modes, *or* poles, *of the matrix A the eigenvalues of A. They are the roots of the characteristic polynomial $\chi_A(z) = \det(zI - A)$ introduced in Definition 2.24.*

Let us now recall the *Cayley-Hamilton theorem* [34].

Theorem 4.10 *If A is a square matrix $n \times n$ and*

$$\chi_A(s) = \det(sI - A) = s^n + a_1 s^{n-1} + \cdots + a_n \tag{4.3}$$

denotes its characteristic polynomial (see Definition 2.24), then we have the following equality between matrices:

$$A^n = -a_1 A^{n-1} - \cdots - a_{n-1} A - a_n I \; . \tag{4.4}$$

In particular, every matrix A^m for $m \geq n$ is a linear combination of $I, A, \ldots, A^{n-1}$.

Routh's criteria (see § 3.4) makes it possible to test the sign of the real parts of the roots by testing the coefficients of a polynomial, which is a simple way to test the asymptotic stability of a matrix without computing its eigenvalues.

4.4 Stability Classification of the Zero Equilibrium for Linear Systems in the Plane

In this section, we consider linear dynamical systems of the form

$$\dot{x} = \begin{pmatrix} a_{11} & a_{12} \\ a_{21} & a_{22} \end{pmatrix} x = Ax \; , \tag{4.5}$$

where A is a nonzero 2×2 real matrix. We classify the stability properties of the zero equilibrium. For this purpose, we need the following result concerning the diagonalization of 2×2 matrices [34]. When they are distinct, the eigenvalues of the matrix A are denoted λ_1 and λ_2.

Proposition 4.11 *If the 2×2 matrix A has two distinct eigenvalues, it is similar to a diagonal matrix on $\mathbb{C}$: if v_1 and v_2 are two (complex) eigenvectors associated with the eigenvalues λ_1 and λ_2, and if $P = (v_1; v_2)$ is the complex matrix corresponding to the basis transformation, we have that*

$$A = P\mathrm{Diag}(\lambda_1, \lambda_2)P^{-1} \text{ where } \mathrm{Diag}(\lambda_1, \lambda_2) = \begin{pmatrix} \lambda_1 & 0 \\ 0 & \lambda_2 \end{pmatrix} . \tag{4.6}$$

The reader can check with Theorem 4.5 that, for 2×2 matrices, asymptotic stability is characterized by the signs of the trace and of the determinant, as follows.

Proposition 4.12 *A 2×2 square matrix A is asymptotically stable if and only if the trace* $\mathrm{Tr}(A)$ *and the determinant* $\mathrm{Det}(A)$ *satisfy*

$$\mathrm{Tr}(A) < 0 \text{ and } \mathrm{Det}(A) > 0 \; . \tag{4.7}$$

One Eigenvalue is Zero

The matrix A being nonzero, we study the case $\lambda_1 = 0$ and $\lambda_2 = \lambda \neq 0$.

In that case, the matrix A has two distinct real eigenvectors v_1 and v_2. If $P = (v_1; v_2)$ denotes the real matrix of basis transformation, we have that $A = P\mathrm{Diag}(0, \lambda)P^{-1}$ by (4.6). With the change of coordinates $z = P^{-1}x$, the linear differential system (4.5) becomes $\dot{z} = \mathrm{Diag}(0, \lambda)z$ and the solutions are:

$$\begin{cases} z_1(t) = z_1(0) \\ z_2(t) = z_2(0)e^{\lambda t}. \end{cases} \tag{4.8}$$

Then, not only the origin, but all the points belonging to the straight line of equation $z_2 = 0$ are equilibrium points. This corresponds to a degenerate case (see Fig. 4.2).

The Eigenvalues are Real, Distinct and with the Same Sign

In that case, the matrix A has two distinct real eigenvectors v_1 and v_2. If $P = (v_1; v_2)$ denotes the real matrix of basis transformation, we have that $A = P\mathrm{Diag}(\lambda_1, \lambda_2)P^{-1}$. With the change of coordinates $z = P^{-1}x$, the linear differential system (4.5) becomes $\dot{z} = \mathrm{Diag}(\lambda_1, \lambda_2)z$, and the solutions are:

$$\begin{cases} z_1(t) = z_1(0)e^{\lambda_1 t} \\ z_2(t) = z_2(0)e^{\lambda_2 t}. \end{cases} \tag{4.9}$$

The origin $z = 0$ is said to be a *knot*, stable or unstable according to the sign of λ_1 and λ_2. In Fig. 4.3, a stable knot is displayed. To draw an unstable knot it suffices to reverse the direction of the arrows.

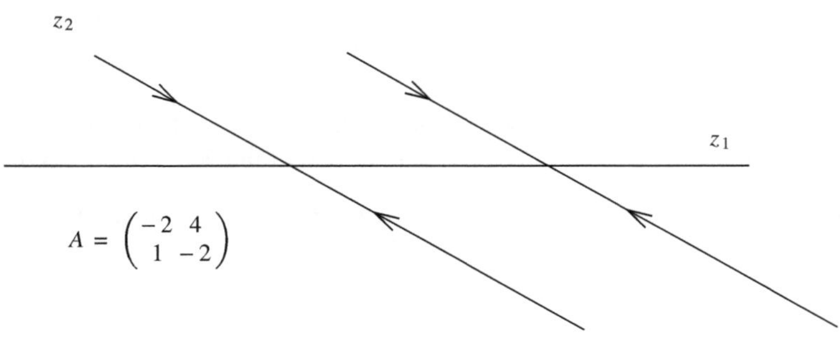

Fig. 4.2 Straight line of equilibrium points

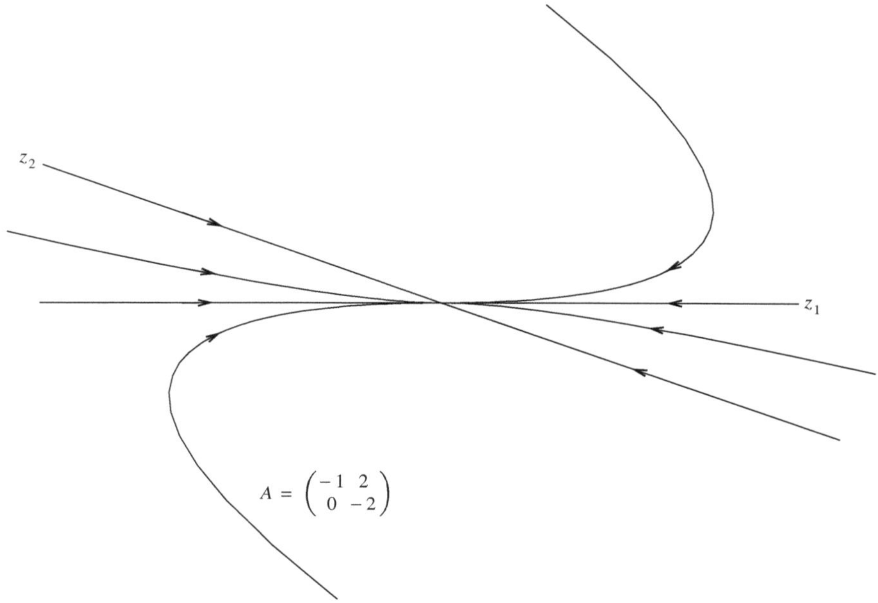

Fig. 4.3 Stable knot

The Eigenvalues are Real, Distinct and with Opposite Sign

The solutions are of the form (4.9). However, since the eigenvalues have an opposite sign, the trajectories are going inside in one direction and outside in the other one. The origin is said to be a *saddle point*. Figure 4.4 shows a saddle point for which $\lambda_1 < 0$ and $\lambda_2 > 0$.

The Eigenvalues are Complex Conjugates

In that case, the matrix A has two distinct complex conjugates vectors v_1 and v_2, associated with the eigenvalues

$$\lambda_1 = \mu + i\theta\ , \quad \lambda_2 = \mu - i\theta\ , \quad \theta \neq 0\ .$$

If $P = (v_1; v_2)$ denotes the complex matrix of basis transformation, we have that $A = P\mathrm{Diag}(\lambda_1, \lambda_2)P^{-1}$. With the change of coordinates $z = P^{-1}x$, the linear differential system (4.5) becomes $\dot{z} = \mathrm{Diag}(\lambda_1, \lambda_2)z$. The complex solution is given by

$$z_1(t) = z_1(0)e^{\mu t}e^{i\theta t}, \quad z_2(t) = z_2(0)e^{\mu t}e^{-i\theta t}, \tag{4.10}$$

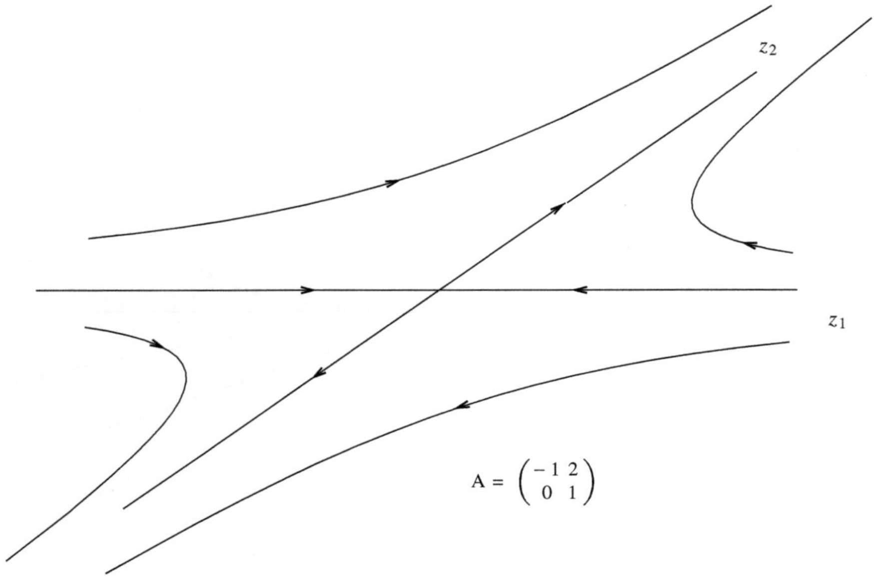

Fig. 4.4 A saddle point

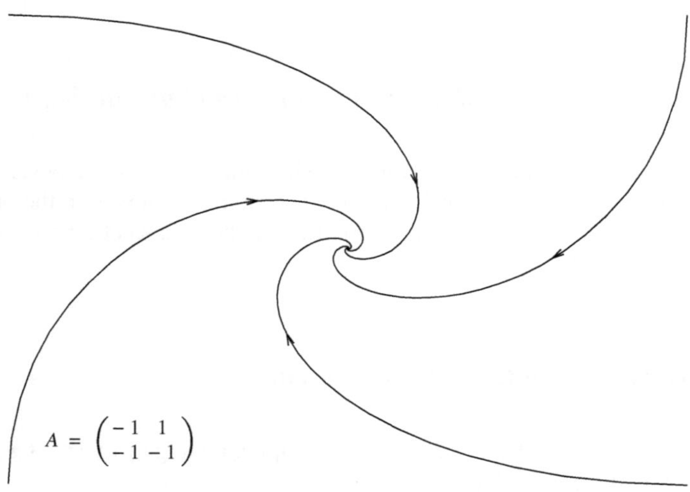

Fig. 4.5 A stable spiral

so that the real solution to (4.5) is a *spiral* around the origin. This trajectory is called an unstable spiral, a *center* or a stable spiral, according to whether μ is strictly positive, zero or strictly negative, as shown in Fig. 4.5.

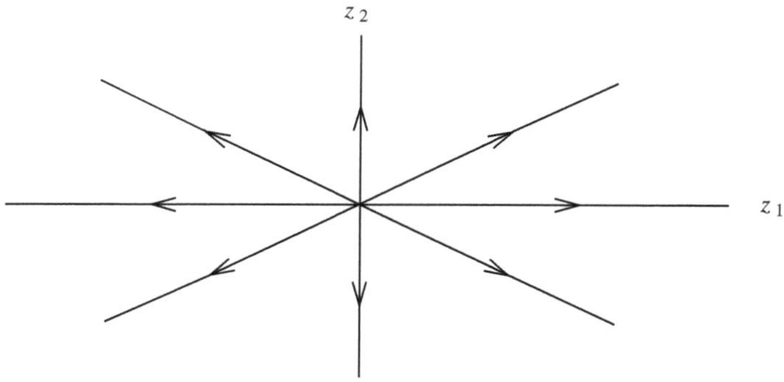

Fig. 4.6 An unstable focus

The Eigenvalues are Equal and A is Diagonalizable

In that case, there is a single real eigenvalue. The solutions have the same form as (4.9). However, since λ_1 and λ_2 are equal, the ratio $z_1(t)/z_2(t)$ remains constant, equal to $z_1(0)/z_2(0)$ and the trajectories are straight lines. The origin is said to be a *focus point*, stable or unstable according to whether λ_1 is strictly negative or strictly positive. Figure 4.6 represents an unstable focus.

The Eigenvalues are Equal and A is not Diagonalizable

There exists a real matrix P of basis transformation such that, from the Jordan decomposition [34], we have that:

$$A = P \begin{pmatrix} \lambda_1 & 0 \\ a & \lambda_1 \end{pmatrix} P^{-1}. \tag{4.11}$$

With the change of coordinates $z = P^{-1}x$, the system (4.5) becomes $\dot{z} = \begin{pmatrix} \lambda_1 & 0 \\ a & \lambda_1 \end{pmatrix} z$, and the solutions are:

$$\begin{cases} z_1(t) = z_1(0)e^{\lambda_1 t} \\ z_2(t) = z_2(0)e^{\lambda_1 t} + atz_1(0)e^{\lambda_1 t}. \end{cases} \tag{4.12}$$

The origin is said to be a *degenerate knot*, stable or unstable according to whether λ_1 is strictly negative or strictly positive. Figure 4.7 represents a stable degenerate knot.

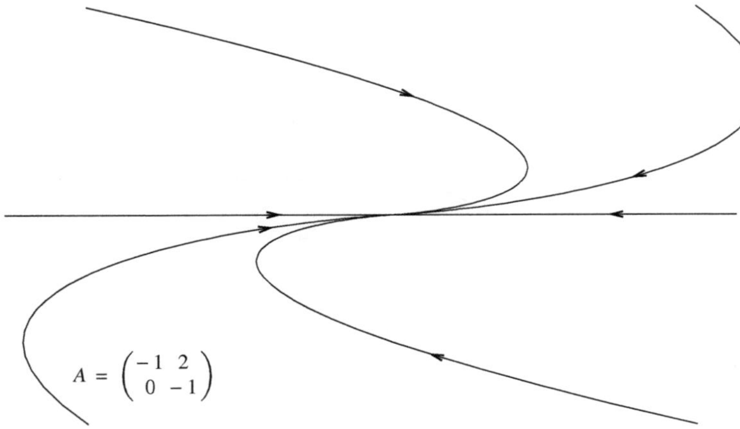

Fig. 4.7 A stable degenerate knot

4.5 Tangent Linear System and Stability

In a proper neighborhood of an equilibrium point, a vector field coincides with the linear term of its Taylor development up to the first-order (recall that, according to Definition 2.10, a vector field is a smooth mapping). In this section, we discuss to what extent the stability of an equilibrium point can be deduced from the stability properties attached to the linear term of the Taylor development, called the *tangent linear system*. For this purpose, we first consider approximation theorems and notions of topological equivalence between flows.

Consider two global flow maps $(\Phi_t)_{t\in\mathbb{R}}$ and $(\Psi_t)_{t\in\mathbb{R}}$: $\mathbb{X} \to \mathbb{X}$ attached to two complete dynamical systems defined over the same phase space $\mathbb{X}$ (see Theorem 2.13 and Definition 2.14).

Definition 4.13 *The flow maps $(\Phi_t)_{t\in\mathbb{R}}$ and $(\Psi_t)_{t\in\mathbb{R}}$ are said to be* equivalent, *if there exists a bijection $\varpi : \mathbb{X} \to \mathbb{X}$, transforming the flow map $(\Phi_t)_{t\in\mathbb{R}}$ into the flow map $(\Psi_t)_{t\in\mathbb{R}}$ in the sense that*

$$\varpi \circ \Phi_t = \Psi_t \circ \varpi \,, \quad \forall t \in \mathbb{R} \,. \tag{4.13}$$

This all amounts to say that $(\Phi_t)_{t\in\mathbb{R}}$ can be transformed into $(\Psi_t)_{t\in\mathbb{R}}$ when changing the coordinates as indicated in Fig. 4.8. By specifying the regularity and the algebraic properties of the transformation ϖ, the following definitions can be introduced.

Definition 4.14 *The flow maps $(\Phi_t)_{t\in\mathbb{R}}$ and $(\Psi_t)_{t\in\mathbb{R}}$ are said to be:*

- linearly equivalent *if the transformation ϖ is an isomorphism (a linear bijection);*
- topologically equivalent *if the transformation ϖ is an homeomorphism (bi-continuous bijection);*

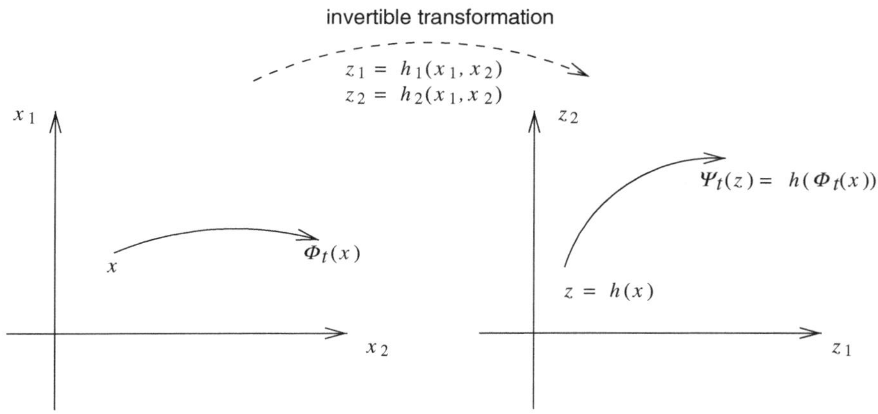

Fig. 4.8 Equivalent flow maps

- differentially equivalent *if the transformation ϖ is a diffeomorphism (differentiable bijection).*

The following results make it possible to obtain a classification of linear dynamical systems. Proofs can be found in [5, 50].

Theorem 4.15 *Let A and F be two real square matrices of dimension n. The linear vector fields associated with the matrices A and F are differentially equivalent if and only if they are linearly equivalent.*

If the matrices A and F have simple eigenvalues (that is, with multiplicity 1 in Definition 2.24), the linear dynamical systems

$$\dot{x} = Ax, \quad x \in \mathbb{R}^n \text{ and } \dot{z} = Fz, \quad z \in \mathbb{R}^n \tag{4.14}$$

are linearly equivalent if and only if the two matrices A and F have the same eigenvalues.

The following theorem is very useful to obtain a topological classification of linear dynamical systems.

Theorem 4.16 *Consider two linear dynamical systems given by* (4.14), *where the matrices A and F have no eigenvalue with zero real part. These two systems are topologically equivalent if and only if the number of eigenvalues of the matrix A having a strictly negative real part (respectively, strictly positive) is equal to the number of eigenvalues of the matrix F having a strictly negative real part (respectively, strictly positive).*

Remark 4.17 Topological equivalence is less restrictive than linear equivalence. ◇

The following notion is essential in the sequel.

Definition 4.18 *Suppose that the dynamical system* (2.35) *has an equilibrium point* x_E. *The* tangent linear system, *also called* first-order approximation, *of the dynamical system* (2.35) *at the equilibrium point* x_E *is the linear dynamical system*

$$\dot{\xi} = A\xi \,, \tag{4.15}$$

where $A = \dfrac{\partial f}{\partial x}(x_E)$ *is the Jacobian matrix associated with the* tangent linear mapping *of* f *at equilibrium point* x_E.

Remark 4.19 In the system of local coordinates $x_1, \ldots, x_n$, the *Jacobian matrix* is made of the partial derivatives of f at x_E. If the vector field f is defined by n differentiable functions $f_1(x_1, \ldots, x_n), \ldots, f_n(x_1, \ldots, x_n)$, we have the expression:

$$\frac{\partial f}{\partial x}(x_E) = \begin{pmatrix} \dfrac{\partial f_1}{\partial x_1}(x_E) & \dfrac{\partial f_1}{\partial x_2}(x_E) & \cdots & \dfrac{\partial f_1}{\partial x_n}(x_E) \\ \vdots & \vdots & \ddots & \vdots \\ \dfrac{\partial f_n}{\partial x_1}(x_E) & \dfrac{\partial f_n}{\partial x_2}(x_E) & \cdots & \dfrac{\partial f_n}{\partial x_n}(x_E) \end{pmatrix} . \tag{4.16}$$

◇

Example 4.20 Computation of the tangent linear approximation of the dynamical model (2.3) describing the evolution of a pendulum without friction.

As discussed in § 4.2, the pendulum dynamics is given by (4.1). The tangent linear approximation in the neighborhood of the equilibrium $x_{inf} = (0, 0)^\top$ is characterized by the matrix

$$A_{inf} = \begin{pmatrix} \dfrac{\partial \dot{\theta}}{\partial \theta} & \dfrac{\partial \dot{\theta}}{\partial \dot{\theta}} \\ -\dfrac{g}{l}\dfrac{\partial \sin\theta}{\partial \theta} & -\dfrac{g}{l}\dfrac{\partial \sin\theta}{\partial \dot{\theta}} \end{pmatrix}_{|\,(\theta,\dot{\theta})=(0,0)} = \begin{pmatrix} 0 & 1 \\ -\frac{g}{l} & 0 \end{pmatrix} . \tag{4.17}$$

The tangent linear approximation in the neighborhood of the equilibrium $x_{sup} = (\pi, 0)^\top$ is characterized by the matrix

$$A_{sup} = \begin{pmatrix} 0 & 1 \\ \frac{g}{l} & 0 \end{pmatrix} . \tag{4.18}$$

△

Linear systems constitute a particularly simple class of dynamical systems, and it is interesting to know to what extent the tangent linear system is similar to the original system in the neighborhood of the origin. From a mathematical point of view, we discuss the topological equivalence between a dynamical system (2.35) and its tangent linear approximation (4.15) around an equilibrium point. An important result is given by the Grobman-Hartman Theorem (see [50] for a proof).

Definition 4.21 *An equilibrium point of a dynamical system is said to be* hyperbolic *if the tangent linear mapping at this point has no eigenvalue with zero real part.*

Theorem 4.22 (Grobman et Hartman) *A nonlinear dynamical system is topologically equivalent to its tangent linear approximation* (4.15) *in the neighborhood of an hyperbolic equilibrium point.*

Therefore, in a sense, the behavior of a nonlinear dynamical system in the neighborhood of an hyperbolic equilibrium point can be deduced from the study of its tangent linear approximation. Regarding stability, the following *perturbation theorem* holds true (this is a corollary of Theorem 4.22).

Theorem 4.23 *Let* x_{E} *be an equilibrium point of the dynamical system* (2.35).

1. *If the equilibrium* x_{E} *is asymptotically stable for the tangent linear system* (4.15), *then it is also asymptotically stable for the original system* (2.35).
2. *If the equilibrium* x_{E} *is unstable for the tangent linear system* (4.15), *then it is also unstable for the original system* (2.35).

If the equilibrium point x_{E} is not hyperbolic, the study of the tangent linear system is generally not sufficient to characterize the stability of the original dynamical system.

Example 4.24 Stability of the pendulum without friction.

The eigenvalues of the matrix A_{inf} in (4.17) are purely imaginary, equal to $\pm i\sqrt{g/l}$. By Proposition 4.4, the matrix A_{inf} is stable, but not asymptotically stable. Therefore, we cannot conclude about the stability of the equilibrium x_{inf} for the original system (4.1) from Theorem 4.23. On the other hand, if viscous damping effects are taken into account by adding an expression of the form $-k\dot{\theta}$, where $k > 0$, in the second term of $f(x)$ in (4.1), the reader can check that the matrix A_{inf} can be written as

$$A_{inf} = \begin{pmatrix} 0 & 1 \\ -\frac{g}{l} & -k \end{pmatrix}.$$

Therefore, the matrix A_{inf} has two eigenvalues with strictly negative real part. As a consequence, the matrix A_{inf} is asymptotically stable. Using Theorem 4.23, we conclude that the equilibrium x_{inf} is now asymptotically stable.

The eigenvalues of the matrix A_{sup} are real with opposite sign, equal to $\pm\sqrt{g/l}$, hence the matrix A_{sup} is unstable. From Theorem 4.23, we deduce that the equilibrium x_{sup} of the original system (4.1) is unstable. △

4.6 Lyapunov Functions and Stability

To remove the ambiguity on the stability property in the nonhyperbolic case, or to evaluate the basin of attraction of an equilibrium point (hence achieving more global stability, see Definition 4.2), so-called *Lyapunov functions* prove quite interesting. Though the stability notions introduced in § 4.2 concern properties of the *flow* of a vector field, the trajectories are generally not known (except for special cases, like the linear case). This is why stability criteria directly depending on the *vector field* are especially relevant. This is the purpose of the stability theory in the sense of *Lyapunov*. Let us first illustrate its mechanical origin on the example of the harmonic oscillator with viscous damping.

Example 4.25 The harmonic oscillator with viscous damping (see Fig. 4.9).

If $\tau(\dot{z})$ represents the term of viscous damping, assumed to be an odd function of the velocity $\dot{z}$, the fundamental principle of dynamics (1.6) gives

$$m\ddot{z} = -kz - \tau(\dot{z}) \ . \tag{4.19}$$

Denoting $x = (z, \dot{z})^\top$, we obtain

$$\frac{dx}{dt} = \begin{pmatrix} \dot{z} \\ \ddot{z} \end{pmatrix} = \begin{pmatrix} \dot{z} \\ \dfrac{-kz - \tau(\dot{z})}{m} \end{pmatrix} = \begin{pmatrix} f^z(z, \dot{z}) \\ f^{\dot{z}}(z, \dot{z}) \end{pmatrix} = f(x) \, . \tag{4.20}$$

The total energy $\mathfrak{E}$ of this physical system is the sum of the kinetic energy and of the potential energy, and we set

$$\mathfrak{V}(z, \dot{z}) = \frac{1}{2}m\dot{z}^2 + \frac{1}{2}kz^2 = \mathfrak{E} \ . \tag{4.21}$$

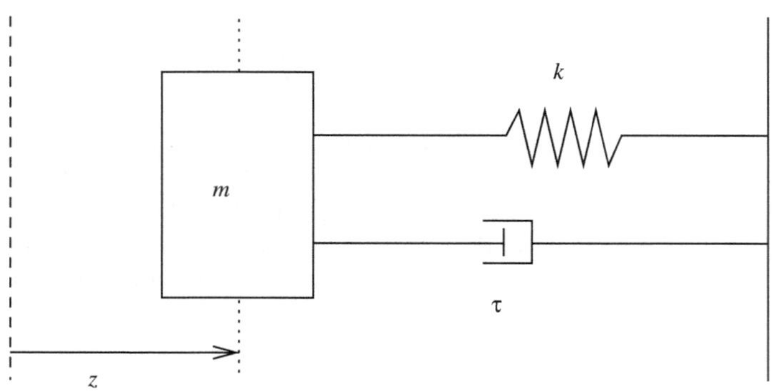

Fig. 4.9 An harmonic oscillator with viscous damping

Let us now study the time evolution of the energy. If $(z_t, \dot{z}_t)$ denotes the solution of the dynamical system (4.20), we have that

$$\begin{aligned}\frac{d\mathfrak{E}}{dt} &= \frac{d}{dt}\left(\frac{1}{2}m\dot{z}^2 + \frac{1}{2}kz^2\right) \text{ by (4.21)} \\ &= m\dot{z}_t\ddot{z}_t + kz_t\dot{z}_t \\ &= \dot{z}_t(m\ddot{z}_t + kz_t) = -\dot{z}_t\tau(\dot{z}_t) \leq 0 \text{ by (4.19)},\end{aligned}$$

since τ is assumed to be odd. Therefore, *the energy decreases along the system's trajectories*. In particular, the trajectories are captured in the domain defined by the level curve $\mathfrak{V}(z, \dot{z}) = \mathfrak{E}(0)$, as shown in Fig. 4.10 where the level curves of the function $\mathfrak{V}$ encircle the equilibrium point $(0, 0)^\top$.

The notion of *Lyapunov function* naturally appears when the computation of the time variation of the energy $\mathfrak{E}$ is made explicit from the relation $\mathfrak{E}(t) = \mathfrak{V}(z_t, \dot{z}_t)$:

$$\begin{aligned}\frac{d\mathfrak{E}}{dt} &= \frac{d}{dt}\mathfrak{V}(z_t, \dot{z}_t) \text{ by (4.21)} \\ &= \frac{\partial\mathfrak{V}}{\partial z}(z_t, \dot{z}_t)\frac{dz_t}{dt} + \frac{\partial\mathfrak{V}}{\partial \dot{z}}(z_t, \dot{z}_t)\frac{d\dot{z}_t}{dt} \\ &= \frac{\partial\mathfrak{V}}{\partial z}(z_t, \dot{z}_t)f^z(z_t, \dot{z}_t) + \frac{\partial\mathfrak{V}}{\partial \dot{z}}(z_t, \dot{z}_t)f^{\dot{z}}(z_t, \dot{z}_t) \text{ by (4.20)} \\ &= \nabla\mathfrak{V}(z_t, \dot{z}_t)\cdot f\left((z_t, \dot{z}_t)^\top\right) \\ &= \dot{\mathfrak{V}}(z_t, \dot{z}_t)\ ,\end{aligned}$$

where $\dot{\mathfrak{V}}$ is defined below in (4.22). The sign of $\dfrac{d\mathfrak{E}}{dt}$ can then be obtained from the sign of the function $\dot{\mathfrak{V}}$ and from the instantaneous values $(z_t, \dot{z}_t)$. △

This example motivates the following definition.

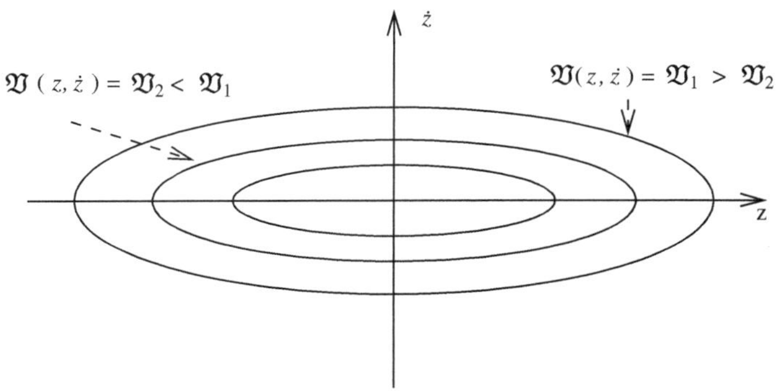

Fig. 4.10 Level curves of the energy function

Definition 4.26 *Let f be a vector field on the phase space $\mathbb{X}$ and $\mathfrak{V} : W \to [0, +\infty[$ be a differentiable function with continuous derivatives. We call* directional derivative *of the function $\mathfrak{V}$ in the direction of the vector field f the function*

$$\dot{\mathfrak{V}}(x) := \nabla\mathfrak{V}(x) \cdot f(x) \ . \tag{4.22}$$

Remark 4.27 If, in the system of coordinates $x_1, \ldots, x_n$, the vector field f is defined by n differentiable functions $f_1(x_1, \ldots, x_n), \ldots, f_n(x_1, \ldots, x_n)$, we have the following expression:

$$\dot{\mathfrak{V}}(x_1, \ldots, x_n) = \sum_{i=1}^{n} \frac{\partial \mathfrak{V}}{\partial x_i}(x_1, \ldots, x_n) f_i(x_1, \ldots, x_n) \ .$$

◇

The expression "directional derivative" comes from the following lemma.

Lemma 4.28 *For all $x \in \mathbb{X}$ and all t such that the flow map $\Phi_t(x)$ of the vector field f is defined, we have that*

$$\frac{d}{dt}\mathfrak{V}\big(\Phi_t(x)\big) = \dot{\mathfrak{V}}\big(\Phi_t(x)\big) \ . \tag{4.23}$$

Proof If $\Phi_t(x)$ is defined, it is also defined in a neighborhood of t by Theorem 2.13. We have that

$$\begin{aligned}
\frac{d}{dt}\mathfrak{V}\big(\Phi_t(x)\big) &= \nabla\mathfrak{V}\big(\Phi_t(x)\big) \cdot \frac{d}{dt}\Phi_t(x) \text{ by composition of derivatives} \\
&= \nabla\mathfrak{V}\big(\Phi_t(x)\big) \cdot f\big(\Phi_t(x)\big) \text{ by (2.40)} \\
&= \dot{\mathfrak{V}}\big(\Phi_t(x)\big) \text{ by (4.22)} \ ,
\end{aligned}$$

which concludes the proof. □

Definition 4.29 *Let x_{E} an equilibrium point of the vector field f. A* Lyapunov function *for f in a neighborhood of x_{E} is a differentiable function $\mathfrak{V}$ with continuous derivatives, defined in a neighborhood of x_{E}, and such that:*

- $\mathfrak{V}(x) > 0$, *except at $x = x_{\mathrm{E}}$ where $\mathfrak{V}(x_{\mathrm{E}}) = 0$;*
- *the directional derivative* (4.22) *of $\mathfrak{V}$ satisfies the inequality $\dot{\mathfrak{V}}(x) \leq 0$.*

The interest of Lyapunov functions comes from the following proposition [28, 46].

Proposition 4.30 *If a Lyapunov function $\mathfrak{V}$ exists for the vector field f in a neighborhood of the equilibrium point x_{E}, then x_{E} is a stable equilibrium for f. In particular, for x in a neighborhood of x_{E}, the flow map $\Phi_t(x)$ is defined for all $t \geq 0$.*

Moreover, if the vector field f and the Lyapunov function $\mathfrak{V}$ are defined in a same neighborhood W of x_E and if the set $\{x \in W \mid \dot{\mathfrak{V}}(x) = 0\}$ does not contain any other invariant subset than the singleton $\{x_E\}$ under the half flow map $(\Phi_t)_{t>0}$ of (2.35), *then x_E is an asymptotically stable equilibrium point of f.*

Proof The proof is derived from [28]. Let W be a neighborhood of x_E on which f and $\mathfrak{V}$ are both defined. Let us consider a compact neighborhood K of x_E included in W and, for $\varepsilon > 0$, let us denote $V_\varepsilon = \{x \in K \mid \mathfrak{V}(x) \leq \varepsilon\}$. We first show that the family $(V_\varepsilon)_{\varepsilon>0}$ constitutes a "fundamental system of neighborhoods of x_E", which makes it possible to consider these neighborhoods in the proofs of stability and of asymptotic stability.

Let $W' \subset K$ be an open neighborhood of x_E. We show that there exists $\varepsilon_0 > 0$ such that $V_{\varepsilon_0} \subset W'$. In fact, since K is a compact neighborhood of x_E, then $K \backslash W' = K \bigcap W'^c$ is closed in K, therefore compact, and the function $\mathfrak{V}$ reaches a minimum $\varepsilon_0 > 0$ on this compact. Therefore, we have that $V_{\varepsilon_0} \subset W'$.

Since the vector field f and the Lyapunov function $\mathfrak{V}$ are defined on a same neighborhood W of x_E, the function $r(t) = \mathfrak{V}\big(\Phi_t(x)\big)$ is well defined for $x \in W$ and t small enough ($0 \leq t \leq t(x)$), and it is decreasing since it satisfies $\dot{r}(t) = \dot{\mathfrak{V}}\big(\Phi_t(x)\big) \leq 0$. Therefore, we have that

$$0 \leq \mathfrak{V}\big(\Phi_t(x)\big) \leq \mathfrak{V}(x) \text{ for } 0 \leq t \leq t(x) . \tag{4.24}$$

We deduce that, for all $\varepsilon > 0$,

$$x \in V_\varepsilon \Rightarrow \Phi_t(x) \in V_\varepsilon . \tag{4.25}$$

Therefore, the trajectory $\Phi_t(x)$ is defined for all $t \geq 0$ since it remains a priori in the compact V_ε. From (4.25) and the fact that the family $(V_\varepsilon)_{\varepsilon>0}$ constitutes a fundamental system of neighborhoods of x_E, we can conclude as to the stability of x_E.

Under the second set of assumptions, let us suppose that the equilibrium point x_E is not attractive. Then there exists $\varepsilon_0 > 0$ and a point $x_0 \in V_{\varepsilon_0}$ such that $\Phi_t(x_0) \not\to x_E$, and then also $\eta > 0$ and a sequence $(t_n)_{n\in\mathbb{N}}$ growing to $+\infty$ such that:

$$\Phi_{t_n}(x_0) \notin V_\eta , \quad \forall n \in \mathbb{N} . \tag{4.26}$$

From (4.24), we deduce that the sequence of points $(\Phi_{t_n}(x_0))_{n\in\mathbb{N}}$ belongs to the compact V_{ε_0}, and therefore admits a sub-sequence (still denoted $\Phi_{t_n}(x_0)$) converging to a point $\overline{x}$. We now show that necessarily $\overline{x} = x_E$, which is inconsistent with (4.26). Indeed, the function $r(t) = \mathfrak{V}\big(\Phi_t(x_0)\big)$ is positive and decreasing and therefore tends to a limit $l \geq 0$. For an arbitrary $s > 0$, the sequences $\mathfrak{V}\big(\Phi_{t_n}(x_0)\big)$ and $\mathfrak{V}\big(\Phi_{s+t_n}(x_0)\big)$ are two subsequences of $r(t)$ which satisfy:

$$\begin{aligned} l &= \lim_{n\to+\infty} \mathfrak{V}\big(\Phi_{t_n}(x_0)\big) = \mathfrak{V}(\overline{x}) \\ l &= \lim_{n\to+\infty} \mathfrak{V}\big(\Phi_{s+t_n}(x_0)\big) = \mathfrak{V}\big(\Phi_s(\overline{x})\big) , \end{aligned}$$

Consequently, $\mathfrak{V}\big(\Phi_s(\overline{x})\big)$ is independent of s and has a zero derivative at s, that is, $\dot{\mathfrak{V}}\big(\Phi_s(\overline{x})\big) = 0$. Then, the half trajectory $\big(\Phi_s(\overline{x})\big)_{s>0}$ belongs to the set $\{x \in W \mid \dot{\mathfrak{V}}(x) = 0\}$. Since the corresponding orbit is clearly invariant under the half flow map $(\Phi_t)_{t\geq 0}$, it is, by assumption, necessarily identical to the equilibrium point x_{E}: $\overline{x} = \Phi_s(\overline{x}) = x_{\mathrm{E}}$. □

A general result about Lyapunov functions is the following *LaSalle Theorem* (we refer the reader to [46, 57] for a proof).

Theorem 4.31 (LaSalle) *Let f be a vector field on the phase space $\mathbb{X}$, and $\mathfrak{V} : W \to [0, +\infty[$ be a differentiable function with continuous derivatives, where W is an open subset of $\mathbb{X}$. Assume that*

- *the function $\mathfrak{V}$ is proper, which means that $\{x \in W \mid \mathfrak{V}(x) \leq R\}$ is compact for all $R > 0$ (this is the case when $\mathfrak{V}$ is defined on $\mathbb{R}^n$ and $\mathfrak{V}(x) \to +\infty$ when $\|x\| \to +\infty$);*
- *the directional derivative* (4.22) *of the function $\mathfrak{V}$ in the direction of the vector field f is negative, that is, $\dot{\mathfrak{V}}(x) = \nabla\mathfrak{V}(x)\cdot f(x) \leq 0$.*

If $\mathbb{I}$ is the largest subset of $\big\{x \in W \mid \dot{\mathfrak{V}}(x) = 0\big\}$ invariant under the flow map (or the half flow map $(\Phi_t)_{t>0}$) of the dynamical system (2.35), *then every trajectory of* (2.35) *is*

1. *included in a compact (and, therefore, bounded);*
2. *attracted by $\mathbb{I}$ in the sense that, for all x in $\mathbb{X}$, we have that $\Phi_t(x) \to_{t\to+\infty} \mathbb{I}$, that is, the distance $dist(\Phi_t(x), \mathbb{I})$ between $\Phi_t(x)$ and $\mathbb{I}$ goes to zero, when t goes to $+\infty$.*

Example 4.32 Consider the Example 4.25 of the harmonic oscillator with viscous damping. By Proposition 4.30, the point $(0, 0)^\top$ is stable since it has been possible to find a Lyapunov function (4.21) for the dynamical system (4.20). Now, to prove asymptotic stability by means of Theorem 4.31, we consider the set

$$\{(z, \dot{z}) \in \mathbb{R}^2 \mid \dot{\mathfrak{V}}(z, \dot{z}) = 0\} = \mathbb{R} \times \{0\} ,$$

and look for invariant subsets under the half flow map $(\Phi_t)_{t>0}$. Setting $\Phi_t(z_0, 0) = (z_t, \dot{z}_t)$, computation shows that

$$z_0 \neq 0 \Rightarrow \frac{d\dot{z}_t}{dt}_{|t=0} = -\frac{k}{m} z_0 \neq 0 .$$

Thus, every trajectory $t \mapsto \Phi_t(z_0, 0)$ leaves $\mathbb{R} \times \{0\}$ for small times $T > 0$, except if $z_0 = 0$. The equilibrium $(0, 0)^\top$ is, of course, invariant. Using LaSalle Theorem 4.31, we conclude that the point $(0, 0)^\top$ is asymptotically stable, since it is the only invariant subset included in the z-axis. △

Finding a Lyapunov function is generally not easy, and it may be helpful to be guided by the nature of the system under study. This is how the total energy of a mechanical system has led to the notion of Lyapunov function.

In the linear case, it is however possible to explicitly compute quadratic Lyapunov functions. This can be useful for the analysis of nonlinear systems in the neighborhood of an equilibrium point, since this quadratic term produces a "first term" of a possible Lyapunov function (see the proof of Proposition 6.17).

Proposition 4.33 *The equilibrium point* $x_{\mathrm{E}} = 0$ *of the linear dynamical system* (2.46) *is asymptotically stable if and only if, for all positive definite matrix* Q*, there exists a positive definite matrix* P *such that:*

$$A^\top P + PA + Q = 0 \,. \tag{4.27}$$

In that case, the function $\mathfrak{V}(x) = x^\top Px$ *is a Lyapunov function for the origin* $x_{\mathrm{E}} = 0$ *of the linear dynamical system* (2.46).

Proof The condition (4.27) is sufficient. Indeed, applying (4.27) with $Q = I$, there exists a positive definite matrix P such that $A^\top P + PA = -I$. Then $\mathfrak{V}(x) = x^\top Px$ defines a Lyapunov function since, by (4.22) and (2.46), it satisfies:

$$\dot{\mathfrak{V}}(x) = x^\top A^\top Px + x^\top PAx = -\|x\|^2 \leq 0 \,.$$

Therefore, by Proposition 4.30, we conclude that the origin $x_{\mathrm{E}} = 0$ of the linear dynamical system (2.46) is asymptotically stable.

Conversely, if the origin $x_{\mathrm{E}} = 0$ of the linear dynamical system (2.46) is asymptotically stable, we know from Theorem 4.5 that all the eigenvalues of the matrix A have strictly negative real part. We deduce, using (2.53), that the following integral is convergent

$$P = \int_0^{+\infty} e^{sA^\top} Qe^{sA}\, \mathrm{d}s \,,$$

and defines a positive definite matrix such that

$$A^\top P + PA = \int_0^{+\infty} \frac{d}{ds}(e^{sA^\top} Qe^{sA})\, \mathrm{d}s = \left[e^{sA^\top} Qe^{sA}\right]_0^{+\infty} = -Q \,.$$

This concludes the proof. □

4.7 Sketch of Stabilization by Linear State Feedback

Most of physical engineering systems have a nonlinear behavior which can be more or less efficiently described through a mathematical model (see Chap. 1). In many cases of mechanical, thermic or biological systems, the control variables appear affinely, which leads to the following state-space representation:

$$\dot{x} = f(x) + g(x)u = f(x) + \sum_{i=1}^{m} g_i(x)u_i \ . \tag{4.28}$$

The *state* vector $x = (x_1, \ldots, x_n)^\top$ belongs to an open set $\mathbb{X} \subset \mathbb{R}^n$, and $u = (u_1, \ldots, u_m)^\top \in \mathbb{U} \subset \mathbb{R}^m$ denotes the vector of *control* variables applied to the system, f is a smooth vector field called *open-loop dynamics vector field* and $g = (g_1, \ldots, g_m)$ is made of the *control vector fields* g_i.

All the results obtained in the sequel could easily be extended to the more general case where $\dot{x} = f(x, u)$.

In many practical cases, we are interested in the behavior of a dynamical system in the neighborhood of an equilibrium point or of a reference trajectory.

Definition 4.34 *A state-control couple* $(x_\mathrm{E}, u_\mathrm{E}) \in \mathbb{X} \times \mathbb{U}$ *or, shortly, a state point* $x_\mathrm{E} \in \mathbb{X}$*, is said to be an* equilibrium *point of the system* (4.28) *if it satisfies*

$$f(x_\mathrm{E}) + g(x_\mathrm{E})u_\mathrm{E} = 0 \ . \tag{4.29}$$

We call tangent linear control system *of the nonlinear system* (4.28) *at* $(x_\mathrm{E}, u_\mathrm{E})$*, the controlled linear dynamical system*

$$\dot{\xi} = A\xi + Bv \ , \tag{4.30}$$

where A *and* B *are the following* $n \times n$ *and* $n \times m$ *matrices*

$$A = \frac{\partial f}{\partial x}(x_\mathrm{E}) + \frac{\partial g}{\partial x}(x_\mathrm{E})u_\mathrm{E} \ , \quad B = g(x_\mathrm{E}) \ . \tag{4.31}$$

REMARKS

- The terminology of *tangent linearized open-loop system* is also used.
- In the more general case where $\dot{x} = f(x, u)$, the equilibrium point x_E satisfies $f(x_\mathrm{E}, u_\mathrm{E}) = 0$ and the matrices A and B are given by:

$$A = \frac{\partial f}{\partial x}(x_\mathrm{E}, u_\mathrm{E}) \ , \quad B = \frac{\partial f}{\partial u}(x_\mathrm{E}, u_\mathrm{E}) \ . \tag{4.32}$$

If we denote

$$\Delta x = x - x_\mathrm{E} \ , \quad \Delta u = u - u_\mathrm{E} \ ,$$

then $A\Delta x + B\Delta u$ is the first term of the Taylor development of $f(x) + g(x)u$. Therefore, the evolution of Δx is governed, at the first-order, by equation (4.30) because[1]

$$\dot{\Delta x} = A\Delta x + B\Delta u + o(\Delta x, \Delta u) .$$

One of the control objectives is to make Δx asymptotically tend to 0, in particular if the matrix A is unstable. For that purpose, one can use the control variables $u_1, \ldots, u_m$ by means of a so-called state feedback.

Definition 4.35 *A* state feedback *is a (smooth) mapping* $\widetilde{u} : \mathbb{X} \to \mathbb{R}^m$ *that associates with every state* $x \in \mathbb{X}$ *a control* $u = \widetilde{u}(x)$.

Remark 4.36 In accordance with Remark 2.4 (see also Remark 3.1), the notations u and x in the equality $u = \widetilde{u}(x)$ correspond here to *vectors*, and not to trajectories. ◇

"Closing" the controlled dynamical system (4.28) with the state feedback $\widetilde{u}$ means that, at each time instant, the control is a function $u = \widetilde{u}(x)$ of the state vector. This procedure yields a classical *closed-loop* dynamical system

$$\dot{x} = f(x) + g(x)\widetilde{u}(x) . \tag{4.33}$$

Its first-order approximation in a neighborhood of the equilibrium x_E is

$$\dot{\xi} = A\xi + B\frac{\partial\widetilde{u}}{\partial x}(x_E)\xi , \tag{4.34}$$

where we assume smoothness of the feedback. Therefore, the dynamics of Δx is given by

$$\dot{\Delta x} = (A - BK)\Delta x + o(\Delta x) , \tag{4.35}$$

where

$$K = -\frac{\partial\widetilde{u}}{\partial x}(x_E) \tag{4.36}$$

is called the *gain matrix*. In many cases (corresponding to "controllability", discussed in § 5.4), a suitable choice of the gain matrix K makes it possible for the closed-loop matrix $A - BK$ to be asymptotically stable and, therefore, to ensure the local asymptotic stability of the original system (4.28) in the neighborhood of the equilibrium point x_E, as shown by the perturbation Theorem 4.23.

In Chap. 5, we focus on elaborating, step by step, such a stabilizing control law. The steps are the following.

- Characterize the stability properties of the open-loop system (4.28).

---

[1] The symbol o corresponds to the *Small-o* notation: $f(z) = o\big(g(z)\big)$ as $z \to z_0$ if and only if $|f(z)|/|g(z)| \to 0$ as z goes to z_0.

- Check if it possible to find a gain matrix K so that the closed-loop matrix $A - BK$ is asymptotically stable. This property is related to the *controllability* property of the linear dynamical system (4.30).
- If the state of the system (4.28) is only partially known, determine if it possible to reconstitute the whole state at each time from the past outputs. This is linked to an *observability* property.

Let us illustrate these notions on the example of the inverted pendulum fixed on a cart moving on an horizontal bench.

Example 4.37 Computation of a linear state feedback stabilizing the tangent linear approximation of the dynamical model (2.3) describing the evolution of a pendulum without friction.

Consider the nonlinear state-space model (2.9) established in § 2.3.1. The tangent linear system in the neighborhood of the equilibrium point given by $z = 0$, $\theta = 0$, $\dot{z} = 0$, $\dot{\theta} = 0$ can be written as

$$\frac{d}{dt}\begin{pmatrix} z \\ \theta \\ \dot{z} \\ \dot{\theta} \end{pmatrix} = A \begin{pmatrix} z \\ \theta \\ \dot{z} \\ \dot{\theta} \end{pmatrix} + Bv \,, \tag{4.37}$$

with

$$\begin{cases} A = \begin{pmatrix} 0 & 0 & 1 & 0 \\ 0 & 0 & 0 & 1 \\ 0 & a & 0 & 0 \\ 0 & b & 0 & 0 \end{pmatrix} , \quad B = \begin{pmatrix} 0 \\ 0 \\ r_1 \\ r_2 \end{pmatrix} , \\ a = -\frac{m}{M} g \,, \quad b = \frac{M+m}{Ml} g \,, \quad r_1 = \frac{1}{M} \,, \quad r_2 = -\frac{1}{Ml} \,. \end{cases} \tag{4.38}$$

The eigenvalues of the open-loop linear dynamical system (4.37), that is, of the matrix A, form the set

$$\mathcal{S}(A) = \left\{ 0, 0, -\sqrt{\frac{M+m}{Ml} g}, \sqrt{\frac{M+m}{Ml} g} \right\} . \tag{4.39}$$

By Proposition 4.4, the open-loop system matrix A is unstable.

Let us suppose that the whole state x is measured, and that we can elaborate a state feedback control law of the form:

$$u = -k_1 z - k_2 \theta - k_3 \dot{z} - k_4 \dot{\theta} \,.$$

The closed-loop linear dynamical system is

$$\dot{x} = (A - BK)x = \begin{pmatrix} 0 & 0 & 1 & 0 \\ 0 & 0 & 0 & 1 \\ -\frac{1}{M}k_1 & -\frac{m}{M}g - \frac{1}{M}k_2 & -\frac{1}{M}k_3 & -\frac{1}{M}k_4 \\ \frac{1}{Ml}k_1 & \frac{M+m}{Ml}g + \frac{1}{Ml}k_2 & \frac{1}{Ml}k_3 & \frac{1}{Ml}k_4 \end{pmatrix} x \,.$$

The eigenvalues of the closed-loop state matrix $(A - BK)$ are determined by the k_i and can be chosen with a strictly negative real part. Unfortunately, in practice it is not always possible to measure the whole state of the system. Let us suppose for example that only the position z and the angle θ are measured. It is then possible, at least formally, to obtain the velocities $\dot{z}$ and $\dot{\theta}$ by time differentiation of the position and angle trajectories. On the other hand, if only the velocities are measured, it is not possible to reconstitute the initial values of the positions $z(0)$ and $\theta(0)$, in which case the system is said to be "unobservable".

In Chap. 5, open-loop criteria are given for linear dynamical systems, making it possible to determine a priori if a system is controllable or observable. In that case, it is possible to elaborate a *stabilizing regulator* and an *asymptotic observer* of the state. △

4.8 Exercises

Exercise 4.8.1 The oscillation movement of a bridge can be described, at first approximation, by the dynamical equation of a spring, that is, by

$$m\ddot{z} = -kz + F,$$

where z denotes the elongation of the bridge with respect to its equilibrium position, and F is the resulting vertical force applied on the bridge. The action of a regiment falling into step on the bridge is represented by a sinusoidal function F with pulsation ω_0, hence satisfying $\ddot{F} = -\omega_0^2 F$.

1. Denoting $\omega_1^2 = k/m$, check that the augmented state $(z, \dot{z}, F, \dot{F})^\top$ satisfies

$$\frac{d}{dt}\begin{pmatrix} z \\ \dot{z} \\ F \\ \dot{F} \end{pmatrix} = \begin{pmatrix} 0 & 1 & 0 & 0 \\ -\omega_1^2 & 0 & 1 & 0 \\ 0 & 0 & 0 & 1 \\ 0 & 0 & -\omega_0^2 & 0 \end{pmatrix} \begin{pmatrix} z \\ \dot{z} \\ F \\ \dot{F} \end{pmatrix}. \tag{4.40}$$

   Let us point out that, for this last system, F and $\dot{F}$ have become internal variables (see § 2.2).
2. Analyze the stability of the point 0 in function of ω_0, ω_1, and comment the results, in particular when $\omega_0 \simeq \omega_1$.

Exercise 4.8.2 Let $\varepsilon > 0$, and consider the dynamical system

$$\begin{cases} \dot{x_1} = x_2 \\ \dot{x_2} = -x_1 - \varepsilon(1 - x_1^2)x_2 \ . \end{cases}$$

1. Studying the function $\mathfrak{V}(x) = \dfrac{1}{2}(x_1^2 + x_2^2)$, show that the equilibrium point $x_E = 0$ is asymptotically stable.
2. Recover this result by using the perturbation Theorem 4.23.

Exercise 4.8.3 Consider the scalar dynamical system:

$$\dot{x} = -x^3 .$$

1. Can we use the perturbation Theorem 4.23 to analyze the stability of the equilibrium point $x_E = 0$?
2. Show that the equilibrium point $x_E = 0$ is asymptotically stable by displaying a Lyapunov function.

Exercise 4.8.4 We have seen, by using Lagrangian mechanics in § 1.4.1, that a mechanical system without constraint and with n degrees of freedom could be described by an equation of the form

$$M(q)\ddot{q} + C(q, \dot{q})\dot{q} + g(q) = Qu \ . \tag{4.41}$$

The constant matrix Q is of rank m, expressing the fact that only m degrees of freedom are directly controlled by an actuator providing a torque or a force $u = (u_1, \dots, u_m)$, with $m \leq n$. The matrix Q can always be chosen of the form:

$$Q = \begin{pmatrix} I_{m\times m} \\ 0_{(n-m)\times m} \end{pmatrix} .$$

Let us also recall that $C(q, \dot{q})\dot{q}$ contains the Coriolis and centrifugal terms, which are quadratic in the velocities $\dot{q}_i$.

1. Let $x_E = (q_E, 0)^\top$ be an equilibrium point of the system (4.41). Compute the corresponding equilibrium control u_E as in (4.29).
2. Show that the open-loop tangent linear system is of the form:

$$\dot{\xi} = \begin{pmatrix} 0 & I \\ -M^{-1}(q_E)\dfrac{\partial g}{\partial q}(q_E) & 0 \end{pmatrix} \xi + \begin{pmatrix} 0 \\ M^{-1}(q_E)Q \end{pmatrix} \upsilon .$$

Exercise 4.8.5 Consider the problem of regulation of the ball rolling on a rail, described in Exercise 2.6.3.

1. Compute the tangent linear system around an equilibrium point $x_E = (\sigma_E, 0, 0, 0)^\top$. The results of Exercise 4.8.4 may be used.
2. Study the stability of the equilibrium point x_E by analyzing the tangent linear system.

Exercise 4.8.6 Consider the nonlinear dynamical system (4.28), where the state vector x and the control u belong to $\mathbb{R}$, and the functions f and g are C^∞. We suppose that $f(0) = 0$, and that the equilibrium point $x_E = 0$ is *globally asymptotically stabilizable*, in the sense that there exists a C^∞ feedback control law $\widetilde{u}(x)$ (as in Definition 4.35) such that $\widetilde{u}(0) = 0$, and that the equilibrium point $x_E = 0$ of the closed-loop system (4.33) is globally asymptotically stable (see Remark 4.3). More precisely, we suppose that a global Lyapunov function associated with the closed-loop system exists such that $\mathfrak{V}(x) \leq 0$ and $\dot{\mathfrak{V}}(x) < 0$ if $x \neq 0$.

We also consider the augmented controlled dynamical system of equation

$$\begin{cases} \dot{x} = f(x) + g(x)z \\ \dot{z} = v \ , \end{cases} \tag{4.42}$$

where v denotes the new control variable. Introducing the function

$$\widetilde{\mathfrak{V}}(x, z) = \mathfrak{V}(x) + \frac{(z - \widetilde{u}(x))^2}{2} \ ,$$

show that the feedback control law

$$\widetilde{v}(x, z) = \frac{\partial \widetilde{u}}{\partial x}(x)(f(x) + g(x)z) - \frac{\partial \mathfrak{V}}{\partial x}(x)g(x) - (z - \widetilde{u}(x))$$

makes the equilibrium point $(x, z)^\top = (0, 0)^\top$ of the controlled dynamical system (4.42), in closed-loop with the feedback law $v = \widetilde{v}(x, z)$, globally asymptotically stable.

Chapter 5
Continuous-Time Linear Dynamical Systems

5.1 Introduction

Here we consider observed and controlled linear dynamical systems in continuous-time. Real systems are rarely linear; they often involve nonlinear behaviors described by nonlinear terms in the dynamical equations of the model. However, the interest of linear dynamical systems is emphasized at the end of this chapter. Indeed, we show that the study of the tangent linear system generally makes it possible to locally stabilize an equilibrium of the nonlinear original system. In contrast to Chaps. 2 and 4, where only two types of variables were considered, namely the state vector and the external input variables, we are now going to take into account *output* or *observed* variables which constitute the components of the state directly measured through sensors. For example, in the case of mechanical systems it is generally easier to measure positions variables (abscissa, angle) than velocities, for precision and cost purposes.

Definition of *continuous-time observed and controlled linear dynamical systems* and examples are given in § 5.2. The so-called "bounded input-bounded state/output" stability is discussed in § 5.3 for controlled and observed linear dynamical systems. Then, we turn to controllability. Intuitively, the controllability property means that the system can be driven from an arbitrary state to another one by means of an open-loop control law. Controllability of linear dynamical systems is characterized in § 5.4, and a linear regulator is displayed that makes it possible to achieve stability. The observability property corresponds, intuitively, to the case when partial knowledge of the state vector makes it possible to reconstruct the entire state. For linear dynamical systems, we provide in § 5.5 a test for observability and we show how an observer can be designed to asymptotically reconstruct the state from linear partial observations. From there, we show in § 5.6 that, for a controllable and observable linear dynamical system, the *estimation-regulation separation principle* holds true: that is, designing a regulator as if the whole state were measured, and then replacing in the regulator law the state by its asymptotic observer does indeed asymptotically stabilize the system. The links between the state-space representation of § 5.2 and the external or

B. d'Andréa-Novel and M. De Lara, *Control Theory for Engineers*,
DOI: 10.1007/978-3-642-34324-7_5,

input-output representation introduced in Chap. 3 are discussed in § 5.7. The results established for the (global) estimation and control of linear dynamical systems are extended to nonlinear dynamical systems in § 5.8, leading to *local* stabilization of a nonlinear dynamical system by linear feedback. The extension to trajectories tracking is discussed in § 5.9. Finally, we sum up in § 5.10 the different steps to elaborate a continuous-time control law, pointing out the practical difficulties and the possible solutions to overcome these difficulties. In particular, the question of sensitivity with respect to parameter uncertainty or control delays is tackled. We conclude by discussing the *stability-precision dilemma.*

5.2 Definitions and Examples

In accordance with Remark 2.4 (see also Remark 3.1), the notations u, x and y in the differential equation (5.1) below correspond to continuous *trajectories* ($t \mapsto u(t)$, $t \mapsto x(t)$ and $t \mapsto y(t)$), and not to vectors. The following differential equation (5.1) has to be understood as equalities between trajectories, that is, for all times t. However, this does not exclude u, x and y to denote *vectors* in the sequel, depending on the context.

Definition 5.1 *An observed and controlled linear dynamical system has the form*

$$\begin{cases} \dot{x} = Ax + Bu \\ y = Cx \,, \end{cases} \tag{5.1}$$

where $u \in \mathbb{R}^m$ represents the m-dimensional vector of control inputs, and $x \in \mathbb{R}^n$ the n-dimensional state vector. The term $y \in \mathbb{R}^p$ denotes the p-dimensional vector of outputs, or observations, or measurements, of the system. The matrix $A_{n\times n}$ is called the state matrix, $B_{n\times m}$ the control matrix, and $C_{p\times n}$ the observation matrix, or output matrix. The space $\mathbb{R}^n$ is the state-space.

Remark 5.2 The *autonomous linear dynamical system*

$$\dot{x} = Ax \tag{5.2}$$

is associated with the linear observed and controlled system (5.1). ◇

Such a system (5.1) is generally obtained through linearization of a controlled nonlinear dynamical system at an equilibrium point (see § 4.7).

Since the system $\dot{x} = Ax$ has the flow map $\Phi_t(x) = e^{tA}x$ by (2.48), it can be easily shown (by the variation of constants method) that the state solution of system (5.1) is given by:

$$x(t) = e^{tA}x_0 + \int_0^t e^{A(t-\tau)}Bu(\tau)\,\mathrm{d}\tau \,. \tag{5.3}$$

Example 5.3 In the Example 4.37 of the inverted pendulum fixed on a cart moving on an horizontal bench, the matrices A and B of the tangent linear system in the neighborhood of the unstable equilibrium are given by (4.38). If only the position z and the angle θ are measured, we obtain the following triplet (where I denotes the 2×2 identity matrix):

$$\begin{cases} A = \begin{pmatrix} 0 & I \\ A_1 & 0 \end{pmatrix} & \text{where } A_1 = \begin{pmatrix} 0 & -\frac{m}{M}g \\ 0 & \frac{M+m}{Ml}g \end{pmatrix} \\ B = \begin{pmatrix} 0 \\ B_1 \end{pmatrix} & \text{where } B_1 = \begin{pmatrix} \frac{1}{M} \\ -\frac{1}{Ml} \end{pmatrix} \\ C = \begin{pmatrix} I & 0 \end{pmatrix}. \end{cases} \tag{5.4}$$

△

Remark 5.4 In some cases, a "direct link" exists between the input and the output of the system, and the following more general state-space representation may be considered

$$\begin{cases} \dot{x} = Ax + Bu \\ y = Cx + Du \ , \end{cases} \tag{5.5}$$

where the matrix $D_{p\times m}$ represents this input-output link. This type of system has already been tackled in Remark 3.22. ◇

Example 5.5 Let us consider the electrical circuit of Fig. 2.12 in Exercise 2.6.2.

The control u is the voltage and the output y is the current intensity. Choosing for state variables the voltage x_1 at the extremities of the capacitor and the current x_2 in the inductor, and applying Kirchhoff's voltage Law recalled in § 2.3.5 provides:

$$\begin{cases} u = x_1 \quad + R_1 C\dot{x}_1 \\ u = L\dot{x}_2 + R_2 x_2 \ . \end{cases}$$

The output y is the sum of the current x_2 in the inductance and the current $i_1 = C\dot{x}_1$ in the capacitor, that is, $y = x_2 - x_1/R + v/R$. This leads to the state-space model (5.5) with state vector $x = (x_1, x_2)^\top$ and matrices:

$$A = \begin{pmatrix} -\dfrac{1}{R_1 C} & 0 \\ 0 & -\dfrac{R_2}{L} \end{pmatrix}, \quad B = \begin{pmatrix} \dfrac{1}{R_1 C} \\ \dfrac{1}{L} \end{pmatrix}$$

$$C = \begin{pmatrix} -\dfrac{1}{R_1} & 1 \end{pmatrix}, \quad D = \frac{1}{R_1} \ .$$

△

Remark 5.6 The existence of the direct input-output link as in (5.5) does not change the system's controllability and observability properties which are developed in the sequel of this chapter. This is why we are mainly interested in the study of systems of the form (5.1). ◇

5.3 Stability of Controlled Systems

In Chap. 4, we have provided notions of stability for stationary autonomous systems, that is, for uncontrolled systems, in the neighborhood of an equilibrium point. We are now going to turn the spotlight onto the notion of "bounded input-bounded state/output" stability in the case of controlled systems.

Definition 5.7 *The linear dynamical system* (5.1) *is said to be* BIBS-stable *if, for all initial condition* x_0 *and for all bounded input trajectory* $(u(t), t \geq 0)$, *the state trajectory* $(x(t), t \geq 0)$ *remains bounded:*

$$\sup_{t \geq 0} \|u(t)\| < +\infty \Rightarrow \sup_{t \geq 0} \|x(t)\| < +\infty . \tag{5.6}$$

The linear dynamical system (5.1) *is said to be* BIBO-stable *if, for all initial condition* x_0 *and for all bounded input trajectory* $(u(t), t \geq 0)$, *the output trajectory* $(y(t), t \geq 0)$ *remains bounded:*

$$\sup_{t \geq 0} \|u(t)\| < +\infty \Rightarrow \sup_{t \geq 0} \|y(t)\| < +\infty . \tag{5.7}$$

Of course, BIBS-stability implies BIBO-stability.

We show that BIBS-stability is connected, as for linear autonomous dynamical systems studied in Chap. 4, to the sign of the real part of the eigenvalues of the state matrix.

Proposition 5.8 *If all the eigenvalues of the matrix* A *have a strictly negative real part, the linear dynamical system* (5.1) *is* BIBS-*stable. If at least one eigenvalue of the matrix* A *has a strictly positive real part, the linear dynamical system* (5.1) *is not* BIBS-*stable.*

Proof If all the eigenvalues of the matrix A have a strictly negative real part, is clear from (5.3) and (2.53) that a bounded control trajectory provides a bounded trajectory for the system.

If at least one eigenvalue of the matrix A has a strictly positive real part, the stationary control trajectory $u(t) = 0$ produces an unbounded state trajectory for a general initial state $x(0)$ (see the proof of Proposition 4.4). □

Remark 5.9 Let us stress the fact that the previous result is not a necessary and sufficient condition for BIBS-stability. Indeed, if $\lambda_1, \ldots, \lambda_r$ are the distinct eigenvalues of the matrix A, the following cases should be considered from Proposition 4.4.

1. If there exists at least one $j = 1, \ldots, r$ such that $\Re(\lambda_j) > 0$, the equilibrium point 0 of the associated autonomous system (5.2) is unstable, and the same holds true for the observed and controlled original system (5.1).
2. If for all $j = 1, \ldots, r$, one has that $\Re(\lambda_j) \leq 0$, then

(a) if $\exists j = 1, \ldots, r$, $\Re(\lambda_j) = 0$ and $\nu(\lambda_j) > 1$, the equilibrium point 0 of the associated autonomous system (5.2), and the same holds true for the observed and controlled original system (5.1);
(b) if $\forall j = 1, \ldots, r$, $(\Re(\lambda_j) = 0 \Rightarrow \nu(\lambda_j) = 1)$, it is not possible to conclude regarding the BIBS-stability property without more information about the matrix B in (5.1).

Therefore, one cannot conclude that a system for which a bounded input provides an unbounded output necessarily has an unstable state matrix A (see Definition 4.1). As an illustration, consider the following *integrator dynamical system*:

$$\dot{x} = u \ .$$

If u is a nonzero constant u_0, the state x is the integral of u, that is, $x(t) = tu_0$. Therefore, $x(t)$ is not bounded. Nevertheless, the system has 0 as eigenvalue and each point constitutes a stable equilibrium of the associated autonomous system, in the sense of Definition 4.1. ◇

Example 5.10 In the case of the inverted pendulum fixed on a cart moving on an horizontal bench (Example 4.37), the state matrix A in (4.38) has a double zero eigenvalue and an eigenvalue with strictly positive real part (see (4.39)). Therefore, the BIBS-stability property is not guaranteed. △

5.4 Controllability. Regulator

Recall that one of the first objectives in Control Theory consists of elaborating control laws that drive a system towards a fixed target. For that purpose, the system should be "controllable." Intuitively, the controllability property means that the system can be driven from an arbitrary state to another one by means of an open-loop control law. On the contrary, the uncontrollability property expresses that some states cannot be reached whatever the control; uncontrollability is often associated with the existence of symmetries (see [33, p. 194]). For linear dynamical systems, we provide a test for controllability, the Kalman controllability criterion. We show that a controllable linear dynamical system can be made equivalent to a canonical form from which it is easy to build a regulator to achieve asymptotic stability.

5.4.1 Controllability

In the case of linear dynamical systems, controllability can be defined as follows.

Definition 5.11 *The linear dynamical system* (5.1) *is said to be controllable if, for all couple* (x_i, x_f) *of state vectors, there exist a finite time* $T \geq 0$ *and a control input* u *defined on* $[0, T]$ *such that, when applying the control* u*, the solution* $x(t)$ *of* (5.1) *with initial condition* $x(0) = x_i$ *satisfies* $x(T) = x_f$.

A very simple algebraic characterization of controllability exists, which is due to Kalman, and is called *Kalman controllability criterion*. The proof of the following Theorem 5.12 consists in the Lemmas 5.15 and 5.16.

Theorem 5.12 *The linear dynamical system* (5.1) *is controllable if and only if the rank of the* controllability matrix

$$\mathcal{C} := \begin{pmatrix} B & AB & \cdots & A^{n-1}B \end{pmatrix} \tag{5.8}$$

is equal to n, the order of the system. In that case, the couple (A, B) *of matrices is also said to be* controllable.

Remark 5.13 Each matrix $A^k B$ having the same dimension $n \times m$ as B, the matrix $\mathcal{C}$ obtained in (5.8) by concatenation of the matrices $B, AB, \ldots, A^{n-1}B$ is of dimension $n \times nm$. ◇

Example 5.14 In the case of the inverted pendulum fixed on a cart moving on an horizontal bench (Example 5.3), the controllability matrix (5.8) can be written, using the notations of (5.4), as

$$\mathcal{C} = \begin{pmatrix} 0 & B_1 & 0 & A_1 B_1 \\ B_1 & 0 & A_1 B_1 & 0 \end{pmatrix}.$$

It can be easily checked that its rank is equal to 4, the order of the system. △

Lemma 5.15 *The linear dynamical system* (5.1) *is controllable if and only if the symmetric positive matrix*

$$P_c(T) = \int_0^T e^{sA} B B^\top e^{sA^\top}\,\mathrm{d}s \tag{5.9}$$

is strictly positive, or equivalently, invertible for at least one $T > 0$.

Proof If the symmetric positive matrix $P_c(T)$ is invertible for at least one $T > 0$, the control law

$$u(s) = B^\top e^{(T-s)A'} P_c(T)^{-1}(x_f - e^{TA}x_i)\ ,\ \ 0 \le s \le T \tag{5.10}$$

drives the initial state $x(0) = x_i$ to $x(T) = x_f$, as shown by the computation of $x(T)$ using formula (5.3).

Conversely, if the symmetric positive matrix $P_c(T)$ is not invertible for a certain $T > 0$, there exists a nonzero vector v such that $v^\top P_c(T)v = 0$. We are going to show that the set of reachable points from 0 is orthogonal to such a vector v, which constitutes an obstruction to controllability. If $x(0) = 0$, we have from (5.3) that

$$v^\top x(t) = \int_0^t v^\top e^{(t-s)A} B u(s)\,\mathrm{d}s\ . \tag{5.11}$$

This last expression is zero because the relation $v^\top P_c(T)v = 0$ can also be written as

$$0 = \int_0^T v^\top e^{sA} B B^\top e^{sA^\top} v \, \mathrm{d}s = \int_0^T \| v^\top e^{sA} B \|^2 \, \mathrm{d}s \, .$$

Therefore, we have that

$$v^\top e^{sA} B = 0 \text{ for } 0 \leq s \leq T \, ,$$

and this remains true beyond T, because $s \mapsto v^\top e^{sA} B$ is an analytical function. As a consequence, we have that

$$\int_0^t v^\top e^{(t-s)A} B u(s) \, \mathrm{d}s = 0 \, , \;\; \forall t \geq 0 \, ,$$

and, from (5.11), that $v^\top x(t) = 0$ for all $t \geq 0$. Therefore, any couple $(0, x_f)$ of state vectors such that $v^\top x_f \neq 0$ cannot comply with the requirements of controllability in Definition 5.11. □

Let us now relate the invertibility property of the symmetric positive matrix $P_c(T)$ in (5.9) to the rank of the controllability matrix $\mathcal{C}$ in (5.8).

Lemma 5.16 *Let $T > 0$. The symmetric positive matrix $P_c(T)$ is definite or, equivalently, invertible if and only if the controllability matrix $\mathcal{C}$ in* (5.8) *has a rank equal to n.*

Proof The following equivalences are satisfied:

$$\begin{aligned}
& P_c(T) \text{ is not invertible} \\
\iff & \exists v \neq 0 \, , \; \int_0^T v^\top e^{sA} B B^\top e^{sA^\top} v \, \mathrm{d}s = 0 \\
\iff & \exists v \neq 0 \, , \; \int_0^T \| v^\top e^{sA} B \|^2 \, \mathrm{d}s = 0 \\
\iff & \exists v \neq 0 \, , \; v^\top e^{sA} B = 0 \, , \; \text{ for } 0 \leq s \leq T \\
\iff & \exists v \neq 0 \, , \; \frac{d^m}{ds^m} (v^\top e^{sA} B)_{s=0} = 0 \, , \; \forall m \in \mathbb{N} \text{ by analyticity of } s \mapsto v^\top e^{sA} B \\
\iff & \exists v \neq 0 \, , \; v^\top A^m B = 0 \, , \; \forall m \in \mathbb{N} \\
\iff & \exists v \neq 0 \, , \; v^\top A^m B = 0 \, , \; \forall m \leq n-1 \text{ by Cayley-Hamilton theorem 4.10} \\
\iff & \exists v \neq 0 \, , \; v^\top (B, AB, \ldots, A^{n-1}B) = 0 \\
\iff & \mathrm{rank}(B, AB, \ldots, A^{n-1}B) < n \, .
\end{aligned}$$

This concludes the proof. □

The proof of the following corollary of Theorem 5.12 is straightforward.

Corollary 5.17 *The following conditions are equivalent.*

1. *The linear dynamical system* (5.1) *is controllable.*
2. *The couple* (A, B) *is controllable.*
3. *Every vector* x *such that* $x^\top B = x^\top AB = \cdots = x^\top A^{n-1}B = 0$ *is zero.*
4. *The symmetric positive matrix* (5.9) *is definite or, equivalently, invertible, for at least one* $T > 0$.

REMARKS

- It can be shown that the set of points reachable from 0 is equal to the vector space generated by the columns of the matrix $(B, AB, \ldots, A^{n-1}B)$ which is also the vector space orthogonal to the kernel of every symmetric matrix $P_c(T)$, for any $T > 0$. All these properties are part of the so-called theory of canonical forms (see for example [30, 33]).
- Notice that the definition of controllability, as well as the Kalman controllability criterion, are totally independent of the notion of observations or outputs of the system, and therefore would have been the same if one had considered the unobserved linear dynamical system in (5.1).

Let us now give a definition of linear dynamical systems equivalence, and introduce the controllable canonical form.

5.4.2 Systems Equivalence. Controllable Canonical Form

We now introduce a canonical structure by means of a change of basis.

Definition 5.18 *Let* (Σ_1) *and* (Σ_2) *be two linear dynamical systems in state-space form having the same number* m *of inputs and the same number* p *of outputs:*

$$(\Sigma_1)\begin{cases} \dot{x} = A_1x + B_1u \\ y = C_1x \end{cases}, \quad (\Sigma_2)\begin{cases} \dot{\xi} = A_2\xi + B_2v \\ \zeta = C_2\xi \ . \end{cases} \tag{5.12}$$

The linear dynamical systems (Σ_1) *and* (Σ_2) *are said to be equivalent (linear dynamical systems) if there exists an invertible square matrix* P *such that if* (u, x, y) *is a solution of* (Σ_1)*, then*

$$v = u \ , \ \xi = Px \ , \ \zeta = y \tag{5.13}$$

is a solution of (Σ_2).

Remark 5.19 The vectors ξ and x in (5.13) have the same dimension, and the following algebraic relations are clearly satisfied:

$$\begin{cases} A_2 = PA_1P^{-1} \\ B_2 = PB_1 \\ C_2 = C_1P^{-1}. \end{cases} \tag{5.14}$$

◇

Proposition 5.20 *If the linear dynamical system* (5.1) *is not controllable, it is equivalent to a system of the form:*

$$\begin{cases} \dot{\xi}_1 = A_1\xi_1 + A_2\xi_2 + B_1u \\ \dot{\xi}_2 = A_3\xi_2 \\ y \ = C_1\xi_1 + C_2\xi_2 \ . \end{cases} \tag{5.15}$$

Proof If the linear dynamical system (5.1) is not controllable, we know by Theorem 5.12 that the rank of the controllability matrix $\mathcal{C}$ in (5.8) is strictly less than n. Then $\mathcal{E}_1 = \text{Im}\mathcal{C}$ is a strict subset of $\mathbb{R}^n$ stable by A. Let $\mathcal{E}_2$ be a supplementary subspace of $\mathcal{E}_1$ in $\mathbb{R}^n$. Choosing a basis of $\mathcal{E}_1$ and of $\mathcal{E}_2$, the form (5.15) is obtained (see for example [44, § 2.4.2]). □

In (5.15), the dynamical evolution of ξ_2 is independent of the control, which constitutes a difficulty if A_3 is unstable, since then $\xi_2(t)$ diverges independently of the control. Indeed, the state matrix of the representation (5.15) has the following triangular form:

$$\begin{pmatrix} A_1 & A_2 \\ 0 & A_3 \end{pmatrix} . \tag{5.16}$$

The eigenvalues of the square matrix (5.16), and consequently of (5.1), are the union of those of A_1 and of A_3.

Definition 5.21 *The eigenvalues of the matrix A_1 in (5.16) are called controllable modes, and the eigenvalues of the matrix A_3 in (5.16) the uncontrollable modes.*

Let us now suppose that the linear dynamical system (5.1) is controllable. We are going to display a suitable vector basis allowing us to establish the link between controllability and a procedure to design a stabilizing regulator.

To simplify, we consider the case of a scalar input, knowing that all the results can be extended to the multi-input case.

Proposition 5.22 *Let A be an $n \times n$ matrix and B an $n \times 1$ matrix. If the couple (A, B) is controllable, there exists an invertible matrix P such that*

$$A_c = PAP^{-1} \ , \ B_c = PB \ , \tag{5.17}$$

with A_c and B_c of the form

$$A_c = \begin{pmatrix} 0 & 1 & 0 & \cdots & 0 & 0 \\ 0 & 0 & 1 & \cdots & 0 & 0 \\ \cdot & \cdot & \cdot & \cdots & \cdot & \cdot \\ \cdot & \cdot & \cdot & \cdots & 1 & 0 \\ 0 & 0 & \cdot & \cdots & 0 & 1 \\ -a_n & -a_{n-1} & \cdot & \cdots & -a_2 & -a_1 \end{pmatrix}, \quad B_c = \begin{pmatrix} 0 \\ 0 \\ \cdot \\ \cdot \\ 0 \\ 1 \end{pmatrix}, \tag{5.18}$$

where the coefficients in the last row of the square matrix A_c are those of the characteristic polynomial (4.3) *of A.*

In other words, the linear dynamical system (5.1) *is equivalent to the following system*

$$\begin{cases} \dot{\xi} = A_c \xi + B_c u \\ y = C P^{-1} \xi \, , \end{cases} \tag{5.19}$$

called the controllable canonical form *of* (5.1).

Proof Let L be the last row of the inverse of the controllability matrix $\mathcal{C}$ given in (5.8):

$$L = (0\,0\,\cdots 0\,1)\,\mathcal{C}^{-1} = (0\,0\,\cdots 0\,1)\left(B \quad AB \cdots A^{n-1}B\right)^{-1}. \tag{5.20}$$

The vector L satisfies

$$LB = LAB = \cdots = LA^{n-2}B = 0 \text{ and } LA^{n-1}B = 1\,. \tag{5.21}$$

Let us define the square matrix

$$P = \begin{pmatrix} L \\ LA \\ \vdots \\ LA^{n-1} \end{pmatrix}.$$

The matrix P is invertible if, and only if, the matrix $P\mathcal{C}$ is invertible, which is equivalent to the property that the rows of $P\mathcal{C}$ are independent. Notice that the matrix $P\mathcal{C}$ is made of the rows $L\mathcal{C}$, $LA\mathcal{C}$, ..., $LA^{n-1}\mathcal{C}$. We have that $L\mathcal{C} = (0\,0\,\cdots 0\,1)$ by (5.20). By (5.21) and Cayley-Hamilton theorem 4.10, we deduce that

$$LA\mathcal{C} = \left(LAB, LA^2B, \cdots, -a_1 LA^{n-1}B - \cdots - a_n LB\right) = (0, 0, \cdots, 1, -a_1)\,.$$

Carrying on in the same way, it can be deduced that the rows of $P\mathcal{C}$ are independent.

Introducing $\xi(t) = Px(t)$, where $x(t)$ satisfies (5.1), we have that $\xi_1 = Lx$ and that, from (5.21),

$$\begin{aligned}
\dot{\xi}_1 &= L\dot{x} = LAx + LBu = LAx = \xi_2 \\
\dot{\xi}_2 &= LA(Ax + Bu) = LA^2x + LABu = LA^2x = \xi_3 \\
\cdots &= \cdots \\
\dot{\xi}_n &= LA^n x + LA^{n-1}Bu = LA^n x + u \ .
\end{aligned}$$

Since $A^n = -a_1 A^{n-1} - \cdots - a_n I$ by Cayley-Hamilton theorem 4.10, we obtain

$$\dot{\xi}_n = -a_n\xi_1 - a_{n-1}\xi_2 - \cdots - a_1\xi_n + u \ ,$$

or, in a matrix form,

$$\dot{\xi} = A_c\xi + B_c u \ , \tag{5.22}$$

where A_c and B_c are given by (5.18). □

Definition 5.23 *The matrices A_c and B_c in (5.18) constitute the controllable canonical form of the matrices A and B, also called the controllable companion form.*

Corollary 5.24 *The coefficients of the characteristic polynomial* (4.3) *of the matrix A are given by the last row of the matrix A_c in* (5.18).

5.4.3 Regulator

In the case of a controllable linear dynamical system, we are now going to discuss how to design a regulator making the zero origin an asymptotically stable equilibrium point.

Definition 5.25 *We call linear state feedback, or linear state regulator, of the linear dynamical system (5.1) a control law (as in Definition 4.35) of the form*

$$\widetilde{u}(x) = -Kx \text{ or } u(t) = -Kx(t) \ , \tag{5.23}$$

where $K_{m\times n}$ is called a counter-reaction gain matrix. Such a state feedback is said to be stabilizing if it ensures the asymptotic stability of the zero equilibrium point of the linear dynamical system (5.1).

Remark 5.26 In this definition, the aim of the regulator is to stabilize the origin. To regulate the system around an arbitrary equilibrium or a trajectory, one rather uses the term *controller* (see § 5.9). ◇

Let us now show that every controllable system can be stabilized by a suitable choice of the regulator gain matrix K in (5.23). More precisely, we can establish the *regulator modes placement* theorem.

Theorem 5.27 *If the couple (A, B) is controllable, a gain matrix K can be chosen to arbitrarily place the modes of the closed-loop matrix $A - BK$. Consequently,* *every controllable linear dynamical system* (5.1) *can be stabilized by a linear state feedback* (5.23).

Proof This theorem also applies for a multi-input system (see for example [44]) but the following proof is restricted to the scalar input case.

Let P be the matrix associated with the change of basis given by Proposition 5.22. We are looking after a control law of the form:

$$u = -K_c P x \ .$$

Setting $\xi = Px$, the closed-loop system (5.22) after feedback can then be written as

$$\dot{\xi} = A_c \xi + B_c u = (A_c - B_c K_c)\xi \ .$$

If $K_c = (k_n \cdots k_1)$, we have that

$$A_c - B_c K_c = \begin{pmatrix} 0 & 1 & 0 & \cdots & 0 & 0 \\ 0 & 0 & 1 & \cdots & \cdot & 0 \\ \cdot & \cdot & \cdot & \cdots & \cdot & \cdot \\ \cdot & \cdot & \cdot & \cdots & 1 & 0 \\ \cdot & \cdot & \cdot & \cdots & 0 & 1 \\ k_n - a_n & k_{n-1} - a_{n-1} & \cdot & \cdots & k_2 - a_2 & k_1 - a_1 \end{pmatrix} , \tag{5.24}$$

and, since the characteristic polynomial (4.3) is invariant by change of basis, we deduce that $\chi_{A_c - B_c K_c}(s) = \chi_{A-BK}(s)$, where $K = K_c P$. Now, from Corollary 5.24, we have that

$$\chi_{A_c - B_c K_c}(s) = s^n + (a_1 - k_1)s^{n-1} + \cdots + (a_n - k_n) \ .$$

It remains to pick up k_1, ..., k_n such that the eigenvalues of $\chi_{A_c - B_c K_c}$ have a strictly negative real part (see Routh criterion in § 3.4). □

From the point of view of practical engineering, the issue is one of choosing where to place these modes in the left half complex plane (see Definition 4.6). The quadratic optimization method, which is discussed in Chap. 7, makes it possible to reduce this problem to the minimization of a cost function, and the resulting control law naturally stabilizes the system.

5.5 Observability. Observer

In § 5.4, we discussed how to stabilize a controllable linear dynamical system by a state feedback. But it could be too expensive to measure the whole state vector of the system, for instance, with position and velocity sensors for a mechanical system. The observability property corresponds, intuitively, to the case when the entire state can be obtained from past history of partial knowledge of the state vector. For linear dynamical systems, we provide a test for observability, the Kalman observability criterion. In that case, we show how an observer can be designed to asymptotically reconstruct the state from partial linear observations.

Definition 5.28 *The linear dynamical system* (5.1) *is said to be observable if, for all vector x_i in $\mathbb{R}^n$, there exists a finite time $T \geq 0$ and a control trajectory $u = (u(t), t \geq 0)$ such that, if $x(0) = x_i$, when applying the input trajectory u, the knowledge of the output trajectory $(y(t), t \in [0, T])$ and of the input trajectory $(u(t), t \in [0, T])$ on the time interval $[0, T]$ makes it possible to determine the initial state x_i.*

As for the controllability property, a simple algebraic observability criterion due to Kalman exists, and is called the *Kalman observability criterion.*

Theorem 5.29 *The linear dynamical system* (5.1) *is observable if and only if the rank of the* observability matrix

$$\mathcal{O} := \begin{pmatrix} C \\ CA \\ \vdots \\ CA^{n-1} \end{pmatrix} \tag{5.25}$$

is equal to n, the order of the system. In that case, the couple (A, C) *is said to be* observable.

Example 5.30 Consider the case of the inverted pendulum fixed on a cart moving on an horizontal bench, discussed in Example 5.3, where only the position z and the angle θ are measured. Using the notation of (5.4), the observability matrix (5.25) can be written as

$$\mathcal{O} = \begin{pmatrix} I & 0 \\ 0 & I \\ A_1 & 0 \\ 0 & A_1 \end{pmatrix}.$$

It is easy to check that its rank is equal to 4, so that the controlled linear dynamical system (4.37)–(4.38), completed with the observation equation

$$y = \begin{pmatrix} I & 0 \end{pmatrix} x$$

is observable. △

The proof of Theorem 5.29 relies on Lemma 5.16 and on the following lemma.

Lemma 5.31 *The linear dynamical system* (5.1) *is observable if and only if the symmetric positive matrix*

$$P_o(T) = \int_0^T e^{sA^\top} C^\top C e^{sA}\,\mathrm{d}s \tag{5.26}$$

is definite or, equivalently, invertible for at least one $T > 0$.

Proof Consider $x_i = x(0)$ as in Definition 5.11. From (5.1) and (5.3), we have that

$$Ce^{tA}x_i = y(t) - \int_0^t Ce^{A(t-\tau)}Bu(\tau)\,\mathrm{d}\tau\ . \tag{5.27}$$

Left multiplying (5.27) by $e^{A^\top}C^\top$ and integrating from 0 to T, we obtain:

$$P_o(T)x_i = \text{function}\,(y(t), u(t), t \in [0, T])\ . \tag{5.28}$$

If, for at least one $T > 0$, the symmetric positive matrix $P_o(T)$ is invertible, the initial state x_i can be deduced from the values of $u(t)$ and $y(t)$ on the time interval $[0, T]$.

Conversely, if for at least one $T > 0$, the symmetric positive matrix $P_o(T)$ is not invertible, it has been proved in Lemma 5.15 that a nonzero vector v exists such that

$$Ce^{tA}v = 0\ ,\ \ \forall t \in [0, T]\ .$$

Therefore, we have that

$$Ce^{tA}(x_i + v) = Ce^{tA}x_i\ ,\ \ \forall t \in [0, T]\ ,$$

and, from (5.27), this means that the states $x_i + v$ and x_i are indistinguishable. Therefore, the linear dynamical system (5.1) is not observable. □

Replacing A by $A^\top$ and B by $C^\top$ in Lemma 5.16, it is easy to show that the invertibility of the symmetric positive matrix $P_o(T)$ in (5.26) is equivalent to the fact that the observability matrix (5.25) has full rank. The proof of the following corollary of Theorem 5.29 is straightforward.

Corollary 5.32 *The following conditions are equivalent.*

1. *The linear dynamical system* (5.1) *is observable.*
2. *The couple (A, C) is observable.*
3. *Every vector x such that $Cx = CAx = \cdots = CA^{n-1}x = 0$ is zero.*
4. *The symmetric positive matrix* (5.26) *is definite or, equivalently, invertible for at least one $T > 0$.*

Remark 5.33 Adding the term Du in the expression (5.27) of the output y does not modify the previous proof. Therefore, the definition and the observability criterion are identical for the linear dynamical systems (5.1) and (5.5). ◇

Proposition 5.34 *If the linear dynamical system* (5.1) *is not observable, it is equivalent to a system of the form:*

$$\begin{cases} \dot{\xi}_1 = A_1\xi_1 + A_2\xi_2 + B_1 u \\ \dot{\xi}_2 = A_3\xi_2 + B_2 u \\ y = C_2\xi_2 \, . \end{cases} \tag{5.29}$$

Proof If the linear dynamical system (5.1) is not observable, the rank of the observability matrix $\mathcal{O}$ in (5.25) is strictly less that n. Then $\mathcal{E}_1 = \mathrm{Ker}\mathcal{O}$ is a strict subspace of $\mathbb{R}^n$ stable by the matrix operator A. Let $\mathcal{E}_2$ be a supplementary subspace of $\mathcal{E}_1$ in $\mathbb{R}^n$. By choosing a suitable basis of $\mathcal{E}_1$ and of $\mathcal{E}_2$, we obtain the form (5.29) (see for example [44, § 2.4.2]). □

Notice that, for the linear dynamical system (5.29), it is impossible to determine ξ_1 from the knowledge of y and u.

Definition 5.35 *The eigenvalues of the matrix* A_1 *in (5.29) are called unobservable modes of* A*, and the eigenvalues of the matrix* A_3 *observable modes of* A.

These definitions are intrinsic, in that the equivalence of linear dynamical systems in Definition 5.18 preserves eigenvalues.

In case the linear dynamical system (5.1) is observable, we now show how we can design a so-called *asymptotic observer*, that is, a dynamical system with state vector $\hat{x}$, driven by the observations y and such that $\hat{x}(t) - x(t)$ goes to 0 when t tends to infinity. The principle of an observer consists in copying the observed system's dynamics and adding a term taking into account the error between the actual output and its estimated value.

Definition 5.36 *A linear asymptotic observer, or Luenberger observer, of the linear dynamical system (5.1) is a state-model of the form*

$$\dot{\hat{x}} = A\hat{x} + Bu - L(C\hat{x} - y) \, , \tag{5.30}$$

with internal variable $\hat{x}$ *and external variables* u *and* y*, where the gain matrix* $L_{n\times p}$ *is such that the solution* $\big(x(t), \hat{x}(t)\big)$ *of the closed system (5.1)–(5.30) satisfies* $\hat{x}(t) - x(t) \to_{t\to+\infty}$ *for every initial conditions* $x(0)$ *and* $\hat{x}(0)$.

As done in § 5.4.2 for the controllability issue, we now display a suitable basis to design a Luenberger observer. To simplify, we tackle the case of a scalar output, knowing that all the results can be extended to the multi-output case [44].

Proposition 5.37 *Let* A *be an* $n \times n$ *matrix and* C *a* $1 \times n$ *matrix. If the couple* (A, C) *is observable, there exists an invertible matrix* P *such that*

$$A_o = P^{-1}AP \, , \; C_o = CP \, , \tag{5.31}$$

with A_o and C_o of the form

$$A_o = \begin{pmatrix} 0 & 0 & \cdots & \cdot & 0 & -a_n \\ 1 & 0 & \cdots & \cdot & 0 & -a_{n-1} \\ 0 & 1 & \cdots & \cdot & \cdot & \cdot \\ \cdot & \cdot & \cdots & \cdot & \cdot & \cdot \\ 0 & 0 & \cdots & 1 & 0 & -a_2 \\ 0 & 0 & \cdots & 0 & 1 & -a_1 \end{pmatrix}, \quad C_o = \begin{pmatrix} 0\, 0 \cdots 0\, 1 \end{pmatrix}, \tag{5.32}$$

where the coefficients in the last column of the square matrix A_O are those of the characteristic polynomial (4.3) *of A.*

In other words, the linear dynamical system (5.1) *is equivalent to the linear dynamical system*

$$\begin{cases} \dot{\xi} = A_o \xi + P^{-1} B u \\ y = C_o \xi \ , \end{cases} \tag{5.33}$$

called the observable canonical form *of the linear dynamical system* (5.1).

Proof The linear dynamical system (5.1) with $p = 1$ being observable by assumption, the observability matrix (5.25) is an $n \times n$ invertible matrix. Let H be the last column of the inverse of the observability matrix $\mathcal{O}$ in (5.25), that is,

$$H = \mathcal{O}^{-1} \begin{pmatrix} 0 \\ \vdots \\ 0 \\ 1 \end{pmatrix} .$$

Then, the vector H is such that

$$CH = CAH = \cdots = CA^{n-2} H = 0 \text{ and } CA^{n-1} H = 1 \ . \tag{5.34}$$

Introducing the square matrix

$$P = \begin{pmatrix} H & AH & \cdots & A^{n-1} H \end{pmatrix},$$

it can be shown, as in the proof of Proposition 5.22, that P is invertible. Defining $x = P\xi$, we have that

$$\begin{cases} \dot{x} = \sum_{i=1}^{n} \dot{\xi}_i A^{i-1} H \\ \dot{x} = Ax + Bu = \sum_{i=1}^{n} \xi_i A^i H + Bu \ . \end{cases}$$

Since $A^n = -a_1 A^{n-1} - \cdots - a_n I$ from Cayley-Hamilton theorem 4.10 and (4.3), the previous equalities give

$$(\dot{\xi}_1 + a_n \xi_n - b_1 u)H + (\dot{\xi}_2 - \xi_1 + a_{n-1}\xi_n - b_2 u)AH + \cdots \\ + (\dot{\xi}_n - \xi_{n-1} + a_1 \xi_n - b_n u)A^{n-1}H = 0 \ ,$$

where B has, in the basis $(H, AH, \cdots, A^{n-1}H)$, been written as

$$B = b_1 H + \cdots + b_n A^{n-1} H = P \begin{pmatrix} b_1 \\ \vdots \\ b_n \end{pmatrix} .$$

Therefore, we obtain another state-space representation of the linear dynamical system (5.1) in the ξ_i-coordinates

$$\begin{cases} \dot{\xi}_1 = -a_n \xi_n + b_1 u \\ \dot{\xi}_2 = \xi_1 - a_{n-1}\xi_n + b_2 u \\ \cdots \ \cdots \\ \dot{\xi}_n = \xi_{n-1} - a_1 \xi_n + b_n u \ , \end{cases}$$

with $y = \xi_n$, since $CP = (0 \cdots 0\ 1)$ by (5.34). We conclude that

$$\begin{cases} \dot{\xi} = A_o \xi + P^{-1} B u \\ y = C_o \xi \ , \end{cases} \tag{5.35}$$

with A_o and C_o given by (5.32). □

Definition 5.38 *The matrices A_o and C_o constitute the observable canonical form of the matrices A and C, also called observable companion form.*

Let us now show that, with a suitable choice of the gain matrix L_o in (5.30), an asymptotic observer can be built for every observable linear dynamical system. More precisely, we are going to prove the following *observer modes placement* theorem.

Theorem 5.39 *If the couple (A, C) is observable, a gain matrix L can be chosen to arbitrarily place the modes of the closed-loop matrix $A - LC$ in* (5.30). *Consequently, a linear asymptotic observer can be elaborated for every observable linear dynamical system* (5.1).

Proof Consider the Luenberger observer of Definition 5.36, and let us define the error vector

$$e = \hat{x} - x \ . \tag{5.36}$$

In the coordinates ξ_i of the observable canonical form (5.33), the dynamical equation of the error e can be written as

$$\dot{e} = (A_o - L_o C_o)e \text{ where } L_o = P^{-1}L = (l_n \cdots l_1)^\top, \tag{5.37}$$

with

$$A_o - L_o C_o = \begin{pmatrix} 0 & 0 & \cdots & 0 & 0 & l_n - a_n \\ 1 & 0 & \cdots & \cdot & 0 & l_{n-1} - a_{n-1} \\ \cdot & \cdot & \cdots & \cdot & \cdot & \cdot \\ \cdot & \cdot & \cdots & 1 & 0 & l_2 - a_2 \\ 0 & 0 & \cdots & 0 & 1 & l_1 - a_1 \end{pmatrix}. \tag{5.38}$$

As for the regulator case discussed in § 5.4.3, it can be seen that a suitable choice of the gain matrix L_o makes it possible to arbitrarily place the modes of the closed-loop matrix $A_o - L_o C_o$, and then to make the error $e(t)$ asymptotically tend to 0. □

Remark 5.40 This theorem still applies for a multi-output system [44]. ◇

Controllability and observability are two dual notions, as specified by the following proposition.

Definition 5.41 *We call dual linear dynamical system of the observed and controlled linear dynamical system (5.1) the observed and controlled linear dynamical system*

$$\begin{cases} \dot{x}_d = A^\top x_d + C^\top u_d \\ y_d = B^\top x_d, \end{cases} \tag{5.39}$$

where the control u_d is a vector of $\mathbb{R}^p$ and the observation y_d a vector of $\mathbb{R}^m$.

Proposition 5.42 *The observed and controlled linear dynamical system* (5.1) *is controllable (respectively, observable) if and only if its dual* (5.39) *is observable (respectively, controllable).*

The proof is straightforward by considering the controllability and observability matrices (5.8) and (5.25) of the two systems.

5.6 Observer-Regulator Synthesis. The Separation Principle

From the discussions in § 5.4 and 5.5, we are able to construct a regulator for a controllable linear dynamical system for which the whole state is known, and an asymptotic observer for an observable linear dynamical system for which only part of the state is measured. Therefore, it seems appealing to design a regulator as if the whole state were measured, and then replace in the regulator control law the state by its asymptotic observer. We are now going to prove that this method indeed stabilizes the system. The property consisting of placing independently the regulator and the observer modes is called the *estimation-regulation separation principle.*

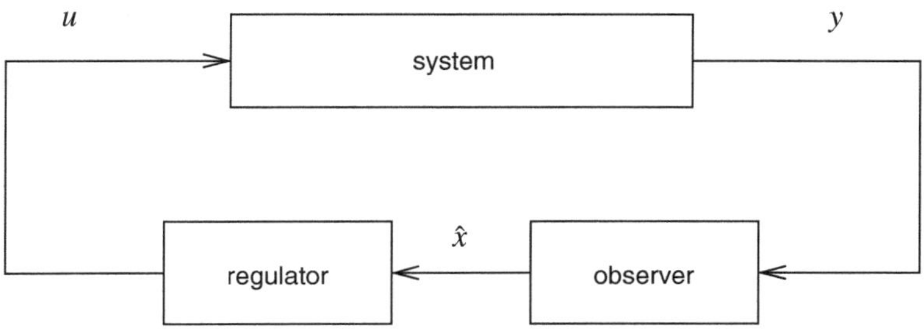

Fig. 5.1 Observer-regulator

Definition 5.43 *We call observer-regulator of the linear dynamical system (5.1) a system with input y, state vector $\hat{x}$ and output u of the form*

$$\begin{cases} \dot{\hat{x}} = A\hat{x} + Bu - L(C\hat{x} - y) \\ u = -K\hat{x} \ , \end{cases} \tag{5.40}$$

as displayed in Fig. 5.1.

The following theorem states the so-called *estimation-regulation separation principle*.

Theorem 5.44 *Assume that the linear dynamical system* (5.1) *is controllable and observable, and let $K_{m\times n}$ and $L_{n\times p}$ be two gain matrices such that $(A - BK)$ and $(A - LC)$ are asymptotically stable matrices. Then, when system* (5.1) *is closed with the observer-regulator* (5.40), *as illustrated in* Fig. 5.1, *we obtain a* closed-loop system, *with state vector $(x, e)^\top = (x, \hat{x} - x)^\top$ and dynamics given by*

$$\begin{cases} \dot{x} = (A - BK)x - BKe \\ \dot{e} = (A - LC)e \ , \end{cases} \tag{5.41}$$

and such that the origin $(x_{\mathrm{E}}, e_{\mathrm{E}})^\top = (0, 0)^\top$ is an asymptotically stable equilibrium point (and this, for all initial condition of the state and of the observer).

Proof Since $y = Cx$ by (5.1) and $u = -K\hat{x}$ by (5.40), the system with state $(x, \hat{x})^\top$ (made from (5.1) and (5.40)) becomes:

$$\begin{cases} \dot{x} = Ax - BK\hat{x} \\ \dot{\hat{x}} = LCx + (A - LC - BK)\hat{x} \ . \end{cases}$$

With the following change of coordinates

$$\begin{pmatrix} x \\ \hat{x} \end{pmatrix} \to \begin{pmatrix} x \\ e \end{pmatrix} = \begin{pmatrix} x \\ \hat{x} - x \end{pmatrix},$$

we obtain (5.41), which can be written in matrix form as

$$\begin{pmatrix} \dot{x} \\ \dot{e} \end{pmatrix} = \begin{pmatrix} A - BK & -BK \\ 0 & A - LC \end{pmatrix} \begin{pmatrix} x \\ e \end{pmatrix}.$$

The modes of this closed-loop system are constituted by the union of the regulator modes—namely, the poles of the matrix $A - BK$—and the observer modes, those of the matrix $A - LC$. Therefore, the closed-loop system (5.41) is stabilized: $(x_E, e_E)^\top = (0, 0)^\top$ is an asymptotically stable equilibrium point. □

Remark 5.45 The *regulator modes* are those of the matrix $A - BK$, and the *observer modes* are the modes of the matrix $A - LC$. ◇

5.7 Links with the Input-Output Representation

We are now going to shed light on the links between the state-space representation (5.1) and the external or input-output representation introduced in Chap. 3.

5.7.1 Impulse Response and Transfer Matrix

By (5.3), the output $y(t)$ of the continuous-time system in state-space form (5.1) has the following expression:

$$y(t) = Ce^{tA}x(0) + \int_0^t Ce^{A(t-\tau)}Bu(\tau)\,\mathrm{d}\tau\ .$$

By considering this relation with zero initial condition ($x(0) = 0$), the notion of impulse response introduced in Definition 3.9 makes it possible to deduce the following proposition.

Proposition 5.46 *The* impulse response *of the time-continuous state-space system* (5.1) *with zero initial condition is given by*

$$h(t) = Ce^{tA}B \text{ if } t \geq 0\ ,\ \ h(t) = 0 \text{ else.} \tag{5.42}$$

This makes it possible to recover the well-known convolution property L2 in § B.1:

$$y(t) = (h \star u)(t)\ ,\ \ \forall t \geq 0\ . \tag{5.43}$$

We also obtain the transfer matrix in the frequency domain.

Proposition 5.47 *The* transfer matrix $H(s)$ *of the time-continuous state-space system* (5.1) *is the following* $p \times m$ *matrix:*

$$H(s) = C(sI - A)^{-1}B \; . \tag{5.44}$$

Proof Let us denote Y, X and U, the respective Laplace transforms of the output, the state and the input, *if they exist*, in the time-continuous state-space system (5.1). Applying the Laplace transform to Eq. (5.1) and using the linearity property and the differentiation Theorem L4 in § B.1, we obtain

$$\begin{cases} sX(s) = AX(s) + BU(s) \\ Y(s) \;\; = CX(s) \; . \end{cases}$$

We deduce that

$$Y(s) = C(sI - A)^{-1}BU(s) \; , \tag{5.45}$$

and, therefore, that the transfer matrix $H(s)$ is indeed given by (5.44). □

We had stated in Chap. 3, without proof, the Proposition 3.16 which can now be easily proved.

Proposition 5.48 *The elements of the transfer matrix* (5.44) *are strictly proper rational functions of* s.

Proof From (5.44), we deduce that $H(s)$ can be written as

$$H(s) = \frac{1}{\det(sI - A)} C \operatorname{cof}(sI - A)^{\top} B \; , \tag{5.46}$$

where $M(s) = \operatorname{cof}(sI - A)^{\top}$ denotes the transpose of the comatrix of $(sI - A)$ [34]. The elements of $M(s)$ are polynomials in s of degree stricIty less than n, the dimension of the square matrix A. Now, the characteristic polynomial $\det(sI - A)$ is of degree n and C and B are constant matrices. As a consequence, the elements of $H(s)$ are strictly proper rational functions. □

Remark 5.49 If a direct link exists between the input and the output, namely if we have that

$$y = Cx + Du \; ,$$

it is straightforward to check that the transfer matrix has the following form:

$$H(s) = C(sI - A)^{-1}B + D \; . \tag{5.47}$$

In that case, by reduction to the same denominator (viz. the characteristic polynomial of A), the elements of the transfer matrix are rational functions, the numerator and denominator of which have the same degree, hence are proper rational functions. Therefore, the system is causal (though not strictly causal) since

$$y(t) = Du(t) + Ce^{tA}x(0) + \int_0^t Ce^{A(t-\tau)}Bu(\tau)\,d\tau,$$

meaning that the output depends on past and present values of the input ($u(\tau)$, $0 \leq \tau \leq t$). ◇

From the definition of the impulse response and the fact that the Laplace transform of a convolution product is a simple product, the following proposition can be deduced.

Proposition 5.50 *The transfer matrix* (5.44) *is the Laplace transform of the impulse response* (5.42) *of the time-continuous linear state-space system* (5.1).

Remark 5.51 By Table B.1 in § B.1, we have that $\mathcal{L}\left[Ce^{tA}B\right](s) = C(sI - A)^{-1}B$. By linearity and injectivity of the Laplace transform, we deduce that

$$e^{tA} = \mathcal{L}^{-1}\left[(sI - A)^{-1}\right](t) . \tag{5.48}$$

◇

5.7.2 From Input-Output Representation to State-Space Representation

When a continuous-time l.c.s. system is described by a transfer matrix $H(s)$ as discussed in § 3.2.2, the so-called *realization* issue is one of computing a state-space model of the form (5.1), given by matrices A, B, C such that:

$$H(s) = C(sI - A)^{-1}B . \tag{5.49}$$

This motivates the following definition.

Definition 5.52 *Every triplet* A, B, C *of matrices satisfying (5.49) is called a continuous-time realization of the transfer matrix* $H(s)$.

As already mentioned in § 5.4.2, the realization of a transfer matrix is not unique. In particular, if the dimension of the state is increased, one can introduce new matrices $\widetilde{A}$, $\widetilde{B}$, $\widetilde{C}$ as follows:

$$\widetilde{A} = \begin{pmatrix} A & 0 \\ \star & \star \end{pmatrix} \quad \widetilde{B} = \begin{pmatrix} B \\ \star \end{pmatrix} \quad \widetilde{C} = \begin{pmatrix} C & 0 \end{pmatrix} .$$

It can easily be checked that they constitute a possible realization of the transfer matrix $H(s)$. This is why we introduce the following notion of realization minimality.

Definition 5.53 *A continuous-time state-space linear dynamical system (5.1) is said to be minimal if its state vector is of minimal dimension in the class of all the systems having the same transfer matrix.*

One can prove the following result (see for example [44, Chap. 6]).

Theorem 5.54 *A continuous-time state-space linear dynamical system* (5.1) *is* minimal *if and only if it is* canonical, *namely controllable and observable.*

A direct computation makes it possible to establish the following result about realization.

Proposition 5.55 *A time-continuous linear scalar system with transfer function*

$$H(s) = b_0 + \frac{\overline{b}_1 s^{n-1} + \cdots + \overline{b}_n}{s^n + a_1 s^{n-1} + \cdots + a_n}$$

has a realization of the following form:

$$\begin{cases} A = \begin{pmatrix} 0 & 1 & 0 \cdots & 0 & 0 \\ 0 & 0 & 1 \cdots & 0 & 0 \\ \cdot & \cdot & \cdots\cdot & \cdot & \cdot \\ \cdot & \cdot & \cdots\cdot & 1 & 0 \\ \cdot & \cdot & \cdots\cdot & 0 & 1 \\ -a_n & -a_{n-1} & \cdot\cdots & -a_2 & -a_1 \end{pmatrix}, \quad B = \begin{pmatrix} 0 \\ 0 \\ \cdot \\ \cdot \\ 0 \\ 1 \end{pmatrix} \\ C = \begin{pmatrix} \overline{b}_n & \cdots & \overline{b}_1 \end{pmatrix}, \quad D = b_0 . \end{cases} \tag{5.50}$$

Remark 5.56 This form has to be compared to the controllable canonical form introduced in Proposition 5.22. Another possible realization is associated with the observable canonical form. ◇

5.7.3 Stability and Poles

We now discuss how stability in § 5.3 is related to the so-called *poles*.

Definition 5.57 *We call poles of the state-space linear dynamical system (5.1) the eigenvalues or modes of the square matrix A.*

Remark 5.58 Definition 5.57 is coherent with Definition 4.9. In the scalar case, from formula (5.46), the *poles of the transfer function* (5.44)—viz. the roots of its denominator—are *poles of the linear dynamical system* (5.1). We prove that they are exactly the system poles if the system is controllable and observable (see Propositions 5.60 and 5.61). ◇

Recall Definition 5.7 saying that the continuous-time linear dynamical system (5.1) is BIBS-*stable* if, for every bounded input, the state remains bounded. With the same arguments than those developed in § 5.3, one can easily deduce the following property.

Proposition 5.59 *The continuous-time linear dynamical system in state-space form* (5.1) *is* BIBS-*stable if its poles have a strictly negative real part.*

Let us emphasize that this is only a sufficient condition: in fact a BIBS-stable or a BIBO-stable system can have a state matrix A which is not asymptotically stable. Indeed, if non stable uncontrollable or unobservable modes exist (see Definitions 5.21 and 5.35), these modes do not appear in the transfer function (see Propositions 5.60 and 5.61).

Proposition 5.60 *The transfer function of the continuous-time linear dynamical system* (5.1) *only depends on its controllable modes.*

Proof From Proposition 5.20, it is known that, in a suitable basis, the system (5.1) can be written as follows:

$$\begin{cases} \dot{x}_1 = A_1 x_1 + A_2 x_2 + B_1 u \\ \dot{x}_2 = A_3 x_2 \\ y \;\; = C_1 x_1 + C_2 x_2 \, . \end{cases}$$

Consequently, the transfer function can be written as

$$H(s) = \begin{pmatrix} C_1 & C_2 \end{pmatrix} \begin{pmatrix} sI - A_1 & -A_2 \\ 0 & sI - A_3 \end{pmatrix}^{-1} \begin{pmatrix} B_1 \\ 0 \end{pmatrix},$$

which gives:

$$H(s) = \begin{pmatrix} C_1 & C_2 \end{pmatrix} \begin{pmatrix} (sI - A_1)^{-1} & \star \\ 0 & (sI - A_3)^{-1} \end{pmatrix}^{-1} \begin{pmatrix} B_1 \\ 0 \end{pmatrix} = C_1 (sI - A_1)^{-1} B_1 \, .$$

This conclude the proof. □

In a similar way, using Proposition 5.34, one can prove the following result.

Proposition 5.61 *The transfer function of the continuous-time linear dynamical system in state-space form* (5.1) *only depends on its observable modes.*

Remark 5.62 Let us recall that the Routh criterion introduced in § 3.4 constitutes a simple algebraic test to check if all the roots of a polynomial have strictly negative real part. ◇

5.8 Local Stabilization of a Nonlinear Dynamical System by Linear Feedback

So far, we have elaborated in § 5.4 and 5.6 a control law allowing, under proper conditions, to stabilize a linear controlled system, that is, to make the zero equilibrium point asymptotically stable. Thanks to the perturbation Theorem 4.23, this process can be extended to the nonlinear case.

We now provide conditions for a linear feedback control law to locally stabilize an equilibrium point (x_E, u_E) of a nonlinear dynamical system of the form (4.28).

Following the notations of § 4.7, consider the nonlinear controlled and observed dynamical system

$$\begin{cases} \dot{x} = f(x) + g(x)u = f(x) + \sum_{i=1}^{m} g_i(x)u_i \\ y = h(x) = \big(h_1(x), \ldots, h_p(x)\big)^\top . \end{cases} \tag{5.51}$$

Let (x_E, u_E) be an equilibrium point, as in Definition 4.34.

Definition 5.63 *We call tangent controlled and observed linear dynamical system of system (5.51) in the neighborhood of the equilibrium point (x_E, u_E), the controlled and observed linear dynamical system*

$$\begin{cases} \dot{\xi} = A\xi + Bv \\ \zeta = C\xi , \end{cases} \tag{5.52}$$

where A, B and C are the following $n \times n$, $n \times m$ and $p \times n$ matrices:

$$A = \frac{\partial f}{\partial x}(x_E) + \frac{\partial g}{\partial x}(x_E)u_E \ , \quad B = g(x_E) \ , \quad C = \frac{\partial h}{\partial x}(x_E) \ . \tag{5.53}$$

Remark 5.64 In the more general case where $\dot{x} = f(x, u)$, we refer the reader to the formulas (4.32). ◇

Theorem 5.65 *Suppose that the tangent controlled and observed linear dynamical system (5.52) of the nonlinear system (5.51) is controllable and observable. Let $K_{m\times n}$ and $L_{n\times p}$ be two gain matrices such that $(A - BK)$ and $(A - LC)$ are asymptotically stable matrices.*

Consider the following linear dynamical system having input y, state $\hat{\xi}$ and output u:

$$\begin{cases} \dfrac{d\widehat{\xi}}{dt} = (A - BK)\widehat{\xi} - L\Big(C\hat{\xi} - \big(y - h(x_E)\big)\Big) \\ u \quad = u_E - K\widehat{\xi} \ . \end{cases} \tag{5.54}$$

Then, the equilibrium point $(x_E, 0)$ of the closed-loop system (5.51)–(5.54), *with state $\big(x, \hat{\xi} - (x - x_E)\big)^\top$, is asymptotically stable.*

Proof Let us set $\Delta x = x - x_E$. From (5.51) and (5.54), we have that

$$\begin{aligned} \frac{d\Delta x}{dt} &= f(x_E + \Delta x) + g(x_E + \Delta x)(u_E - K\widehat{\xi}) \\ &= A\Delta x + B(-K\widehat{\xi}) + \varepsilon_1(\Delta x, -K\widehat{\xi}) \end{aligned}$$

$$= (A - BK)\Delta x - BK(\widehat{\xi} - \Delta x) + \varepsilon_1(\Delta x, -K\widehat{\xi}) ,$$

where ε_1 is negligible with respect to its arguments in the neighborhood of 0.

If we define $\Delta e(t) = \widehat{\xi}(t) - \Delta x(t)$, then $(\Delta x(t),\ \Delta e(t))^\top$ satisfies the differential equation

$$\frac{d}{dt}\begin{pmatrix}\Delta x(t)\\ \Delta e(t)\end{pmatrix} = \begin{pmatrix}A - BK & -BK\\ 0 & A - LC\end{pmatrix}\begin{pmatrix}\Delta x(t)\\ \Delta e(t)\end{pmatrix} + \begin{pmatrix}\varepsilon_1(\Delta x, -K(\Delta e + \Delta x))\\ -\varepsilon_1\big(\Delta x, -K(\Delta e + \Delta x)\big) + L\varepsilon_2(\Delta x)\end{pmatrix},$$

where ε_2 is negligible with respect to its arguments in the neighborhood of 0.

The asymptotic stability of the tangent linear dynamical system makes it possible to conclude the proof, thanks to the perturbation Theorem 4.23. □

In contrast to the case of a controlled and observed linear dynamical system, especially Theorem 5.44, the linear feedback control law of Theorem 5.65 is stabilizing only when the initial state $x(0)$ belongs to a neighborhood of the equilibrium point x_{E} and $\hat{\xi}(0)$ to a neighborhood of the origin.

5.9 Tracking Reference Trajectories

Up to now, we have tackled the problem of stabilizing the origin of a linear dynamical system in Theorem 5.44 and, locally, an equilibrium point of a nonlinear dynamical system in Theorem 5.65. We now discuss how the state feedback laws elaborated in § 5.4 and 5.6 can be extended to stabilize an equilibrium point of a linear dynamical system and a reference trajectory of a nonlinear dynamical system.

5.9.1 Stabilization of an Equilibrium Point of a Linear Dynamical System

Consider the linear dynamical system:

$$\dot{x} = Ax + Bu . \tag{5.55}$$

As discussed in § 5.4.3, if the couple (A, B) is controllable, the origin 0 can be asymptotically stabilized through a regulator of the form $u = -Kx$ where the counter-reaction gain matrix K can be chosen such that the matrix $A - BK$ is asymptotically stable.

Now, let us consider an equilibrium point x_{E} different from the origin, namely a point satisfying

$$Ax_{\mathrm{E}} + Bu_{\mathrm{E}} = 0 ,$$

where u_E denotes the equilibrium value of the control. If we are now interested in stabilizing the linear dynamical system (5.55) around the equilibrium point x_E, we can consider the following *controller*:

$$u = -Kx + v \ . \tag{5.56}$$

The control v, called *auxiliary control*, is chosen as follows

$$v = u_E + Kx_E \ , \tag{5.57}$$

and is such that the dynamics of the closed-loop error $\Delta x = x - x_E$ can be written as

$$\dot{\Delta x} = A\Delta x + B(u - u_E) = (A - BK)\Delta x \ .$$

This implies the asymptotic convergence of Δx towards 0, and therefore the convergence towards x_E of the state vector x of the linear dynamical system (5.55).

5.9.2 Stabilization of a Slowly Varying Trajectory

Consider the state-space dynamical system with state vector x and input u:

$$\dot{x} = f(x, u) \ . \tag{5.58}$$

Let $x_c(t)$ be a smooth trajectory solution of

$$\dot{x}_c(t) = f\big(x_c(t), u_c(t)\big) \ ,$$

and consider the controller

$$u = -Kx + v \ \text{ with } \ v(t) = u_c(t) + Kx_c(t) \ . \tag{5.59}$$

If we denote

$$\overline{f}_t(x) = f\Big(x, u_c(t) - K\big(x - x_c(t)\big)\Big)$$

the vector field of system (5.58) in closed-loop with (5.59), the dynamical equation of the linear dynamical system tangent to the trajectory can be written as

$$\dot{\Delta x} = F(t)\Delta x \ \text{ with } \ F(t) = \frac{\partial \overline{f}_t}{\partial x}(x_c(t)) \ . \tag{5.60}$$

Let us prove the following proposition.

Proposition 5.66 *Suppose that the following assumptions hold true.*

1. *The orbit $\{x_c(t) \mid t \geq 0\}$ of the reference trajectory $t \mapsto x_c(t)$ is included in a compact set.*
2. *The trajectory $t \mapsto x_c(t)$ is slowly varying, viz. $\|\dot{x}_c(t)\| < \varepsilon$, where $\varepsilon > 0$ is sufficiently small.*
3. *For each "frozen" $(t, x_c(t))$, the origin is an asymptotically stable equilibrium point of the stationary linear dynamical system obtained by considering* (5.60) *at a frozen $(t, x_c(t))$.*

Then, the solution $x(t)$ of $\dot{x} = \overline{f}_t(x)$ asymptotically tends to the slowly varying reference trajectory $x_c(t)$.

Proof Defining the closed-loop error vector $\Delta x = x - x_c$, we can write:

$$\dot{\Delta x} = F(t)\Delta x + b(t, \Delta x) \text{ with } \| b(t, \Delta x) \| < k \| \Delta x \|^2 , \ \forall t \geq 0 . \tag{5.61}$$

Let us first consider the time-varying tangent linear dynamical system (5.60). Applying to this system a stability result for slowly varying systems (see [64, Chap. 5]), it can be shown that the origin is an asymptotically stable equilibrium point of system (5.60). In a second step, an approximation result (see for example [38]) makes it possible to conclude to the asymptotic stability of the origin for the nonlinear dynamical system (5.61). This concludes the proof. □

Remark 5.67 The feedback control law given by (5.59) is such that the same gain matrix K is supposed to asymptotically stabilize any frozen equilibrium point $x_c(t)$. By continuity of the eigenvalues of the matrix $F(t)$ in (5.60) with respect to t, this assumption is quite realistic if the compact where the trajectory $x_c(t)$ is defined for all t is sufficiently small. ◇

On the other hand, let us point out that, if the trajectory $x_c(t)$ is such that the matrix $F(t)$ is constant (independent of t), a same gain matrix K can clearly stabilize every frozen point $x_c(t)$. This is precisely what happens when one wants to stabilize a mechanical system along reference trajectories for which the inertia matrix remains invariant. The reader can easily check this property from the general model of mechanical systems discussed in § 1.4.1.

Example 5.68 Consider the example of the inverted pendulum fixed on a cart moving on an horizontal bench described in § 2.3.1. We are interested in making the cart follow a slowly varying reference trajectory, while the pendulum stays at its unstable vertical position. The reference trajectory is of the form

$$x_c(t) = (z_c(t), 0, \dot{z_c}(t), 0)^\top , \tag{5.62}$$

with $\|\dot{z_c}(t)\|$ small. The equations (2.9) are independent of the position z_c of the cart and the same holds true for the tangent linear dynamical system at any frozen equilibrium $x_c(t)$, so that a same gain matrix stabilizes any equilibrium at a frozen time t. △

5.9.3 Stabilization of Any State Trajectory

The tangent linear dynamical system of (5.58) along a state trajectory $t \mapsto x_c(t)$ is given by

$$\dot{\xi} = A_c(t)\xi + B_c(t)v ,$$

with

$$A_c(t) = \frac{\partial f}{\partial x}\big(x_c(t), u_c(t)\big) , \quad B_c(t) = \frac{\partial f}{\partial u}\big(x_c(t), u_c(t)\big) .$$

This is a time-varying linear dynamical system for which the controllability and observability notions may be extended (see for example [44]), but no general stabilization result is available. Nevertheless, one can try to elaborate so-called optimal feedback laws, which are generally time-varying but not necessarily stabilizing (see § 7.2.1). The same kind of difficulty arises in estimation problems around any trajectory (see § 7.3.1).

Remark 5.69 We have tackled the problem of stabilizing trajectories by means of state feedback laws. However, the results can be extended to the case where the state is not totally measured, and where it is possible to design an asymptotic observer. ◇

5.10 Practical Set Up. Stability-Precision Dilemma

We are now going to sum up the different steps to elaborate a continuous-time control law, pointing out the practical difficulties and the possible solutions to overcome them. In particular, the question of sensitivity with respect to parameter uncertainty or control delays is tackled. We conclude by discussing the so-called *stability-precision dilemma*.

5.10.1 Steps for the Elaboration of a Control Law

In the rest of the book, the *control synthesis* or *control design* consists of the following steps.

1. Define the system and the set point around which we want the system to be regulated.
2. Determine the available control inputs and the possible observations or measurements on the system, as well as the disturbances.
3. Elaborate a mathematical model describing the system dynamics.
4. In the case of a state-space model, define the set point as an equilibrium point and compute the tangent linear dynamical system at this point.
5. Check the controllability property of the tangent linear dynamical system.

6. Choose the outputs (measurements) of the system and check the observability property of the tangent linear dynamical system.
7. If the algebraic controllability and observability properties are satisfied, compute the controller and observer gain matrices.
8. Write, implement and test the control algorithm.

Some questions and difficulties arise at each step.

1. *Define the system and the set point around which we want the system to be regulated.*
   Different systems may be defined according to different precision levels, and this can lead to different mathematical models.
2. *Determine the available control inputs and the possible observations or measurements on the system as well as the disturbances*
   In some systems, the input and output variables are determined without ambiguity, whereas in others, the situation is more open, in which case various options to drive the system can be screened, as well as physical limits of the actuators (saturations, maximal range, etc.). In the same way, different available measurements can be discussed, as well as physical limits of the corresponding sensors (precision, noise, etc.). Concerning the disturbances, the problem is to identify their source and try to reduce them.
3. *Elaborate a mathematical model describing the system dynamics.*
   Taking into account the two first steps and the nature of the system, some variables emerge, allowing to describe the system's behavior through a mathematical model. Such a model should combine fidelity to the original system and simplicity. This is a difficult compromise since precision of the model is on a par with complexity.
   Some approximations are generally necessary to obtain a satisfying state-space model. Since many models are possible candidates to describe a system, the choice of one of them depend on some control objectives. We refer the reader to the distinction made between *knowledge models* and *control models* in the Foreword.
   From a practical point of view, it is important to take into account in the system's equations the physical parameters which are pertinent to describe the system's behavior and whose values are approximatively known. The consequence of parameter uncertainty is studied below.
   Concerning the disturbances, a way to take them into account by a probabilistic approach is casted at Chap. 7. We distinguish between parameter uncertainty, which are mainly constant (stationary), and disturbances which are generally time-varying.
4. *In the case of a state-space model, define the set point as an equilibrium point and compute the tangent linear dynamical system at this point.*
   At this stage, we have elaborated a model of the form

$$\dot{x} = f(x, u, p, q) \tag{5.63}$$

where x denotes the state vector, u the input vector, p the vector of disturbances and q the vector of parameters.
An equilibrium state x_E mathematically satisfies the system of equations $f(x_E, u_E, p_E, q_E) = 0$ where u_E is the equilibrium input, and p_E and q_E are the corresponding disturbances and parameters. If it is possible, the analytical resolution can be realized using formal languages. If it is not possible, the resolution is done numerically. The dependency of the equilibrium x_E with respect to (u_E, p_E, q_E) is important in the two cases (analytical and numerical) and is studied below.
The same problems arise for the estimation of the matrices

$$A = \frac{\partial f}{\partial x}(x_E, u_E, p_E, q_E) \text{ and } B = \frac{\partial f}{\partial u}(x_E, u_E, p_E, q_E) \; .$$

5. *Check the controllability property of the tangent linear dynamical system.*
   The problem is to evaluate analytically or numerically the rank of the $n \times mn$ controllability matrix $(B, AB, \ldots, A^{n-1}B)$.
6. *Choose the outputs (measurements) of the system and check the observability property of the tangent linear dynamical system.*
   The choice of measurements (or output functions) $y = h(x)$ may be guided first by the aim of having an observable couple (A, C), where $C = \dfrac{\partial h}{\partial x}(x_E)$.
7. *If the algebraic controllability and observability properties are satisfied, compute the controller and observer gain matrices.*
   Computing a gain matrix K such that $A - BK$ is asymptotically stable using the method described in the proof of Theorem 5.27 is delicate and not enough robust from a numerical point of view (matrix inversion). Moreover, how to choose the modes of the matrix $A - BK$ is a widely open question, and this choice determines the stability and precision performances of the closed-loop system, as is shown below. We highlight a robust pole placement method in Chap. 7.
8. *Write, implement and test the control algorithm.*
   To apply the control algorithm, the digital character of computers forces us to tackle sampling issues. In fact, measurements are generally obtained at discrete-time instants and the control is feed into the system also at discrete times and maintained constant on regular time intervals. This problem is studied in Chap. 6.

5.10.2 Sensitivity to Model Parameter Uncertainty: Precision

Consider a state-space model of the form

$$\dot{x} = f(x, u, q) \tag{5.64}$$

where x denotes the state vector, u the input vector and q the vector of parameters (the disturbances are not considered here).

For certain parameters values $q = \overline{q}$, the model (5.64) constitutes a good representation of the system dynamics. These values are approximately known and all the computations on the system (5.64) are realized with values $q = q_{\mathrm{E}} \approx \overline{q}$. We are interested here to study the impact of such uncertainties on the control design described above.

The following proposition is a consequence of the implicit function Theorem [10].

Proposition 5.70 *Suppose that the mapping $f(x, u, q)$ is C^∞ in its arguments and that, for a given value q_{E} of the parameter, an equilibrium point $(x_{\mathrm{E}}, u_{\mathrm{E}})$ exists, namely $f(x_{\mathrm{E}}, u_{\mathrm{E}}, q_{\mathrm{E}}) = 0$. If the matrix*

$$A(x_{\mathrm{E}}, u_{\mathrm{E}}, q_{\mathrm{E}}) = \frac{\partial f}{\partial x}(x_{\mathrm{E}}, u_{\mathrm{E}}, q_{\mathrm{E}})$$

is invertible, then

1. *there exist a neighborhood $W_{x_{\mathrm{E}}}$ of x_{E}, a neighborhood $W_{u_{\mathrm{E}}}$ of u_{E} and a neighborhood $W_{q_{\mathrm{E}}}$ of q_{E}, such that for all (u, q) in $W_{u_{\mathrm{E}}} \times W_{q_{\mathrm{E}}}$, there exists a unique x in $W_{x_{\mathrm{E}}}$ such that $f(x, u, q) = 0$;*
2. *if we denote $x = \phi(u, q)$ the solution hereabove, the mapping ϕ is C^∞ and we have that*
$$\frac{\partial \phi}{\partial q}(u_{\mathrm{E}}, q_{\mathrm{E}}) = -\big(A(x_{\mathrm{E}}, u_{\mathrm{E}}, q_{\mathrm{E}})\big)^{-1}\frac{\partial f}{\partial q}(x_{\mathrm{E}}, u_{\mathrm{E}}, q_{\mathrm{E}}) \; .$$

The practical consequences of this proposition are the following.

If the matrix $A(x_{\mathrm{E}}, u_{\mathrm{E}}, q_{\mathrm{E}})$ is invertible and if the "true" value $q = \overline{q}$ of the parameters is sufficiently close to q_{E}, then there exists an equilibrium point $\overline{x}$ of (5.64) sufficiently close to x_{E}. Let us point out that other equilibrium points of (5.64) may exist for $q = \overline{q}$, but, in that case, they are necessarily outside the neighborhood $W_{x_{\mathrm{E}}}$.

For a closed-loop system, the previous remarks make it possible to emphasize the link between precision and the choice of the counter-reaction gain matrix of the regulator. Suppose that the tangent linear dynamical system of (5.64) (see Definition 5.63) is controllable and let K be a matrix such that, with obvious notations, $A(x_{\mathrm{E}}, u_{\mathrm{E}}, q_{\mathrm{E}}) - B(x_{\mathrm{E}}, u_{\mathrm{E}}, q_{\mathrm{E}})K$ is asymptotically stable. Let us apply to the system (5.64), for $q = \overline{q}$, the following linear state feedback control law

$$u = u_{\mathrm{E}} - K(x - x_{\mathrm{E}}) \; . \tag{5.65}$$

The issue of *precision* is: do the trajectories of the closed-loop system converge or not towards a state close to x_{E}?

The closed-loop vector-field can be written as

$$f_{cl}(x, q) = f(x, u_{\mathrm{E}} - K(x - x_{\mathrm{E}}), q) \; ,$$

and we set

$$A_{cl}(x,q) = \frac{\partial f_{cl}}{\partial x}(x,q) = \frac{\partial f}{\partial x}(x, u_{\mathrm{E}} - K(x - x_{\mathrm{E}}), q) - \frac{\partial f}{\partial u}(x, u_{\mathrm{E}} - K(x - x_{\mathrm{E}}), q)K.$$

For $x = x_{\mathrm{E}}$ and $q = q_{\mathrm{E}}$, the matrix $A_{cl}(x_{\mathrm{E}}, q_{\mathrm{E}})$ is the state matrix of the tangent linear dynamical system to the nonlinear system (5.64) with feedback law (5.65), and it is equal to

$$A_{cl}(x_{\mathrm{E}}, q_{\mathrm{E}}) = A(x_{\mathrm{E}}, u_{\mathrm{E}}, q_{\mathrm{E}}) - B(x_{\mathrm{E}}, u_{\mathrm{E}}, q_{\mathrm{E}})K,$$

which is asymptotically stable, hence invertible. Therefore, if $\overline{q} \approx q_{\mathrm{E}}$, there exists an equilibrium point $\overline{x}$ of the closed-loop system

$$\dot{x} = f_{cl}(x, \overline{q}) = f(x, u_{\mathrm{E}} - K(x - x_{\mathrm{E}}), \overline{q}) \tag{5.66}$$

sufficiently close to the equilibrium point x_{E}. Moreover, by continuity of all the terms defining the matrix $A_{cl}(x,q)$, we have that $A_{cl}(x_{\mathrm{E}}, q_{\mathrm{E}}) \approx A_{cl}(\overline{x}, \overline{q})$. From Proposition 2.28, as soon as $\overline{q}$ is sufficiently close to q_{E}, the eigenvalues of the matrix $A_{cl}(\overline{x}, \overline{q})$ are sufficiently close to those of the matrix $A_{cl}(x_{\mathrm{E}}, q_{\mathrm{E}})$. Then, at least locally, the state trajectory of the closed-loop system (5.66) converges towards $\overline{x}$ and the first-order approximation can be written as

$$\overline{x} - x_{\mathrm{E}} \approx -(A_{cl}(x_{\mathrm{E}}, q_{\mathrm{E}}))^{-1} \frac{\partial f_{cl}}{\partial q}(x_{\mathrm{E}}, q_{\mathrm{E}})(\overline{q} - q_{\mathrm{E}}) .$$

Thus, to have a good precision, the term $\big(A_{cl}\big(x_{\mathrm{E}}, q_{\mathrm{E}}\big)\big)^{-1}$ should be "small". In the scalar case, the higher the gain K the smaller the real number $\big(A_{cl}\big(x_{\mathrm{E}}, q_{\mathrm{E}}\big)\big)^{-1}$.

5.10.3 Sensitivity to Input Delay: Stability

Up to now, we have supposed that the state feedback law $u = u_{\mathrm{E}} - K(x - x_{\mathrm{E}})$ is applied instantaneously, in the sense that the control $u(t)$ applied at time t is a function of the state $x(t)$ at the same time instant. But, in practice, a delay τ may possibly exist, which can be taken into account in the feedback law as follows:

$$u(t) = u_{\mathrm{E}} - K(x(t-\tau) - x_{\mathrm{E}}) .$$

For $\overline{q} = q_{\mathrm{E}}$, the closed-loop system (5.66) can be written as

$$\begin{aligned} \dot{x}(t) &= f(x(t), u_{\mathrm{E}} - K(x(t-\tau) - x_{\mathrm{E}}), q_{\mathrm{E}}) \\ &= A(x_{\mathrm{E}}, u_{\mathrm{E}}, q_{\mathrm{E}})(x(t) - x_{\mathrm{E}}) + o(x(t) - x_{\mathrm{E}}) \\ &\quad - B(x_{\mathrm{E}}, u_{\mathrm{E}}, q_{\mathrm{E}})K(x(t-\tau) - x_{\mathrm{E}}) + o(x(t-\tau) - x_{\mathrm{E}}) . \end{aligned}$$

This last equation cannot be analyzed by means of the mathematical tools developed up to now. Nevertheless, one can write an expansion $x(t-\tau) = x(t) - \tau\dot{x}(t) + \cdots$, which gives

$$\dot{x}(t) = (A(x_\mathrm{E}, u_\mathrm{E}, q_\mathrm{E}) - B(x_\mathrm{E}, u_\mathrm{E}, q_\mathrm{E})K)(x(t) - x_\mathrm{E}) + \tau B(x_\mathrm{E}, u_\mathrm{E}, q_\mathrm{E})K\dot{x}(t) + r(t),$$

where the $r(t)$ contains all the neglected terms. This last equation can be written as

$$\dot{x}(t) = (I - \tau B(x_\mathrm{E}, u_\mathrm{E}, q_\mathrm{E})K)^{-1}(A(x_\mathrm{E}, u_\mathrm{E}, q_\mathrm{E}) - B(x_\mathrm{E}, u_\mathrm{E}, q_\mathrm{E})K)(x(t) - x_\mathrm{E}) + r(t),$$

where it can be seen that, if the gain K is too high, the system dynamics can be completely disturbed by the term $(I - \tau B(x_\mathrm{E}, u_\mathrm{E}, q_\mathrm{E})K)^{-1}$. Indeed, in the scalar case, we can write

$$\dot{x}(t) = \frac{A(x_\mathrm{E}, u_\mathrm{E}, q_\mathrm{E}) - B(x_\mathrm{E}, u_\mathrm{E}, q_\mathrm{E})K}{1 - \tau B(x_\mathrm{E}, u_\mathrm{E}, q_\mathrm{E})K}(x(t) - x_\mathrm{E}) + r(t) \; .$$

Now, if $B(x_\mathrm{E}, u_\mathrm{E}, q_\mathrm{E})K \gg 1$, we have that $A(x_\mathrm{E}, u_\mathrm{E}, q_\mathrm{E}) - B(x_\mathrm{E}, u_\mathrm{E}, q_\mathrm{E})K < 0$, but also that $1 - \tau B(x_\mathrm{E}, u_\mathrm{E}, q_\mathrm{E})K < 0$, which means that the stability property is lost!

Thus, ensuring a good precision requires, on the one hand, that

$$A(x_\mathrm{E}, u_\mathrm{E}, q_\mathrm{E}) - B(x_\mathrm{E}, u_\mathrm{E}, q_\mathrm{E})K \ll -1 \; ,$$

as explained in § 5.10.2, whereas, on the other hand, to preserve stability, we need that

$$B(x_\mathrm{E}, u_\mathrm{E}, q_\mathrm{E})K \ll 1 \; ,$$

and these two conditions are inconsistent. This fact is the so-called *stability-precision dilemma*, that makes difficult the pole placement of the closed-loop system. In Chap. 7, we illuminate on a possible way to solve this question. Another illustration of the stability-precision dilemma has been given in § 3.7, in the context of the frequency-domain approach.

5.11 Exercises

Exercise 5.11.1 Consider the chemical process studied in Exercise 2.6.1. The objective of the control is to regulate the concentration of output product at a constant value $\overline{c}$ associated with a constant flow $\overline{\varphi}$ corresponding to an equilibrium point with constant input flows $\overline{\varphi}_1$, $\overline{\varphi}_2$ and a constant volume $\overline{V}$ of solution in the vessel.

We define the state vector

$$x = (x_1, x_2)^\top \; , \; x_1(t) = V(t) - \overline{V} \; , \; x_2 = c(t) - \overline{c} \; ,$$

and the control input vector

$$u = (u_1, u_2)^\top, \quad u_1(t) = \varphi_1(t) - \overline{\varphi}_1, \quad u_2(t) = \varphi_2(t) - \overline{\varphi}_2.$$

1. Determine the steady-state equations, and show that the open-loop tangent linear dynamical system of (2.54), given in Definition 5.63, is

$$\dot{x} = \begin{pmatrix} -\alpha & 0 \\ 0 & -2\alpha \end{pmatrix} x + \begin{pmatrix} 1 & 1 \\ \beta_1 & \beta_2 \end{pmatrix} u, \tag{5.67}$$

   where

$$\alpha = \frac{\overline{\varphi}}{2\overline{V}}, \quad \beta_1 = \frac{c_1 - \overline{c}}{\overline{V}}, \quad \beta_2 = \frac{c_2 - \overline{c}}{\overline{V}}.$$

2. Determine the stability and the nature of the equilibrium point of the associated classical dynamical system, that is, (5.67) with $u = 0$. Plot its trajectories in a neighborhood of the origin.
3. Study the BIBS-stability of system (5.67).
4. Assuming that the output of the system has the expression

$$y = \begin{pmatrix} \alpha & 0 \\ 0 & 1 \end{pmatrix} x,$$

   study the observability property of the tangent linear dynamical system.

Exercise 5.11.2 Consider the electrical circuit studied in Exercise 2.6.2, where the current is assumed to be measured (see Fig. 2.12 of the corresponding exercise).

1. Write the observation equation.
2. For which conditions the system is not controllable? not observable?
3. Give a physical interpretation of these conditions, in terms of the system time constants.

Exercise 5.11.3 Consider the regulation problem of a ball rolling on a rail discussed in Exercises 2.6.3 and 4.8.5.
Study the controllability and observability properties of the tangent linear dynamical system.

Exercise 5.11.4 Consider the state-space linear dynamical system (5.1).

1. Recall the general form of equivalent state-space representations of this system.
2. Show that two state-space equivalent systems have the same transfer matrix.

Exercise 5.11.5 Suppose that the linear dynamical system (5.1) is not observable. Show that its transfer matrix only depends on the observable modes.

Chapter 6
Discrete-Time Linear Dynamical Systems

6.1 Introduction

The digital character of computers incites us now to turn the spotlight onto *sampled* control systems, where the control is fed into the system only at discrete times. These times are often regularly spaced, and the *sampling period* is the interval between two consecutive times. This interval is bounded below by the time required by the processor to compute the new control.

The control structure of a sampled control system can be described as follows (see Fig. 6.1):

- a microprocessor computes the controls and carries out the sampling;
- D/A (digital/analogical) converters transform digital (discrete) quantities into analogical (continuous) quantities;
- sensors provide measurements, generally analogical;
- A/D converters (analogical/digital) sample at a given frequency the analogical quantities coming from the sensors to convert them into digital quantities.

Linear dynamical systems in discrete-time have analogies with their counterparts in continuous-time. However, they possess some specific concepts and properties that are highlighted in this chapter. In § 6.2, we describe how a continuous-time linear dynamical system can be sampled, yielding an *exact discretized* discrete-time linear dynamical system. Then, we investigate the stability of discrete-time classical dynamical systems in § 6.3, and that of controlled discrete-time linear dynamical systems in § 6.4. Controllability and regulator design are tackled in § 6.5, observability and observer design in § 6.6, making it possible to focus on observer-regulator synthesis and on the separation principle in § 6.7. The choice of the sampling period is discussed in § 6.8. The links between controlled discrete-time linear dynamical systems and the input-output representation are the subject of § 6.9. Finally, we show how a control law in *discrete*-time can locally stabilize a nonlinear *continuous*-time dynamical system in § 6.10.

B. d'Andréa-Novel and M. De Lara, *Control Theory for Engineers*,
DOI: 10.1007/978-3-642-34324-7_6, © Springer-Verlag Berlin Heidelberg 2013

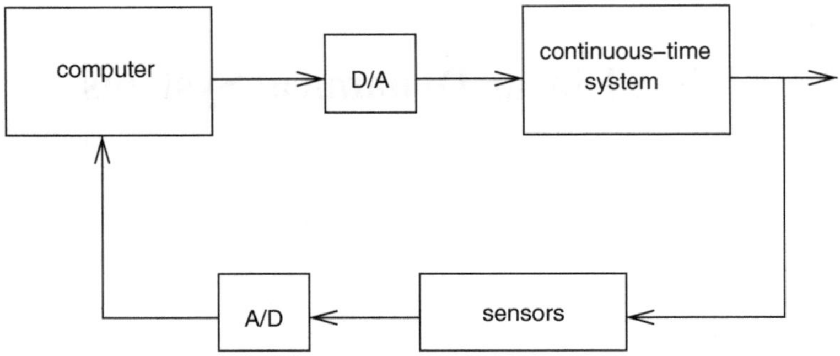

Fig. 6.1 Digital control process

6.2 Exact Discretization of a Continuous-Time Linear Dynamical System

Consider a continuous-time linear dynamical system of the form

$$\begin{cases} \dot{x} = Ax + Bu \\ y = Cx \ , \end{cases} \tag{6.1}$$

where u represents the m-dimensional vector of control, x the n-dimensional vector of state and y the p-dimensional vector of output. If x_0 is the initial state of (6.1), we know that

$$x(t) = e^{At} x_0 + \int_0^t e^{A(t-\tau)} Bu(\tau)\, \mathrm{d}\tau \ . \tag{6.2}$$

Definition 6.1 *Let* $\Delta T > 0$ *denote the* sampling period. *A D/A converter is called* zero-order hold *(denoted by* ZOH*)* at period ΔT, *or* at frequency $1/\Delta T$, *if it transforms a continuous-time signal* $(u(t), t \in \mathbb{R})$ *into the stepwise signal, constant on the interval* $[k\Delta T, (k+1)\Delta T[$ *having value* $u(k\Delta T)$, *for* $k \in \mathbb{Z}$.

There are converters of higher order, approximating more finely the signal on the sampling interval, but they are less used than the ZOH.

Sampling the linear dynamical system (6.1) with a zero-order hold amounts to maintaining the input u constant on the interval $[k\Delta T, (k+1)\Delta T[$. Taking $x_k = x(k\Delta T)$ as initial state of the continuous-time linear dynamical system (6.1), and applying the control $u_k = u(k\Delta T)$, one can, by the formula (6.2), compute the next state $x\big((k+1)\Delta T\big)$. This is how we obtain the following *discretization Theorem*.

Theorem 6.2 *The* linear exact discretized dynamical system *of the continuous-time linear dynamical system* (6.1) *sampled by a* ZOH *at frequency* $1/\Delta T$ *is given by the* discrete-time state representation

$$\begin{cases} x_{k+1} = A_d x_k + B_d u_k \\ y_k = C x_k \ , \end{cases} \tag{6.3}$$

for $k \in \mathbb{N}$, *with*

$$A_d = e^{A\Delta T} \ , \ B_d = \left(\int_0^{\Delta T} e^{At} \, \mathrm{d}t \right) B \ . \tag{6.4}$$

Proof We set

$$x_k = x(k\Delta T) \ , \ \forall k \in \mathbb{N} \ , \tag{6.5}$$

where $t \mapsto x(t)$ solves (6.1). By (6.2), the expression of $x_{k+1} = x\big((k+1)\Delta T\big)$ is of the form:

$$x_{k+1} = e^{A\Delta T} x_k + \int_{k\Delta T}^{(k+1)\Delta T} e^{A\left((k+1)\Delta T - \tau\right)} Bu(\tau) \, \mathrm{d}\tau \ .$$

The input u being constant, equal to u_k on the interval $[k\Delta T, (k+1)\Delta T[$, we obtain the following expression, after the change of variable $\tau \to -s + (k+1)\Delta T$ in the integral:

$$x_{k+1} = e^{A\Delta T} x_k + \left(\int_0^{\Delta T} e^{At} \, \mathrm{d}t \right) Bu_k \ .$$

Setting

$$y_k = y(k\Delta T) \ , \ \forall k \in \mathbb{N} \ , \tag{6.6}$$

where $t \mapsto y(t)$ solves (6.1), we easily deduce that $y_k = Cx_k$. This ends the proof. □

The expression of the state x_n of the linear dynamical system (6.3) at the discrete time n is

$$x_n = A_d^n x_0 + \sum_{k=0}^{n-1} A_d^{n-1-k} B_d u_k \ , \tag{6.7}$$

where x_0 denotes the initial state.

Remarks

- If the matrix A in (6.3) is invertible, one can show that the matrix B_d in (6.4) is given by

$$B_d = A^{-1}(e^{A\Delta T} - I)B \ . \tag{6.8}$$

- With the continuous-time controlled and observed linear dynamical system (6.3), we associate the *discrete-time classical linear dynamical system*

$$x_{k+1} = A_d x_k \ . \tag{6.9}$$

Example 6.3 Consider the continuous-time linear dynamical system:

$$\begin{cases} \begin{pmatrix} \dot{x}_1 \\ \dot{x}_2 \end{pmatrix} = \begin{pmatrix} 0 & 1 \\ -1 & 0 \end{pmatrix} \begin{pmatrix} x_1 \\ x_2 \end{pmatrix} + \begin{pmatrix} 0 \\ 1 \end{pmatrix} u \\ y = \begin{pmatrix} 1 & 0 \end{pmatrix} \begin{pmatrix} x_1 \\ x_2 \end{pmatrix} . \end{cases} \tag{6.10}$$

The exact discretized with a ZOH at sampling period ΔT is given by

$$\begin{cases} \begin{pmatrix} x_{1k+1} \\ x_{2k+1} \end{pmatrix} = \begin{pmatrix} \cos \Delta T & \sin \Delta T \\ -\sin \Delta T & \cos \Delta T \end{pmatrix} \begin{pmatrix} x_{1k} \\ x_{2k} \end{pmatrix} + \begin{pmatrix} 1 - \cos \Delta T \\ \sin \Delta T \end{pmatrix} u \\ y = \begin{pmatrix} 1 & 0 \end{pmatrix} \begin{pmatrix} x_{1k} \\ x_{2k} \end{pmatrix} . \end{cases} \tag{6.11}$$

△

Example 6.4 In the case of the inverted pendulum on a cart, discussed in § 2.3.1 and in Example 4.37, applying the discretization Theorem 6.2 to (4.38) yields the exact discretized linear dynamical system characterized by the matrices

$$A_d = \begin{pmatrix} 1 & \frac{a}{\omega^2}(\cosh(\omega \Delta T) - 1) & \Delta T & -\frac{a}{\omega^3}(\omega \Delta T - \sinh(\omega \Delta T)) \\ 0 & \cosh(\omega \Delta T) & 0 & \frac{1}{\omega}\sinh(\omega \Delta T) \\ 0 & \frac{a}{\omega}\sinh(\omega \Delta T) & 1 & \frac{a}{\omega^2}(\cosh(\omega \Delta T) - 1) \\ 0 & \omega \sinh(\omega \Delta T) & 0 & \cosh(\omega \Delta T) \end{pmatrix} \tag{6.12}$$

$$B_d = \frac{1}{Ml} \begin{pmatrix} l\frac{\Delta T^2}{2} + \frac{a}{\omega^4}\left(1 + \omega^2 \frac{\Delta T^2}{2} - \cosh(\omega \Delta T)\right) \\ \frac{1}{\omega^2}(\cosh(\omega \Delta T) - 1) \\ l \Delta T + \frac{a}{\omega^3}\left(\omega \Delta T - \sinh(\omega \Delta T)\right) \\ -\frac{1}{\omega}\sinh(\omega \Delta T) \end{pmatrix} , \tag{6.13}$$

where we have set

$$a = -\frac{m}{M} g , \quad \omega = \sqrt{\frac{M + m}{Ml} g} .$$

△

Remark 6.5 If, instead of the continuous-time linear dynamical system (6.1), we consider the more general case where the control u is directly linked to the output y by $y = Cx + Du$ as in (5.5), the exact discretized is

$$\begin{cases} x_{k+1} = A_d x_k + B_d u_k \\ y_k = C x_k + D u_k . \end{cases} \tag{6.14}$$

As in the continuous-time case, the existence of the direct link Du from the control u to the output y does not change the properties of controllability and of observability of the system that we develop. This is why we focus on linear dynamical systems of the form (6.3). ◇

6.3 Stability of Discrete-Time Classical Dynamical Systems

Consider a *discrete-time classical dynamical system*

$$x_{k+1} = \Phi(x_k) \ , \ \ \forall k \in \mathbb{N} \ , \tag{6.15}$$

where $\Phi : \mathbb{X} \to \mathbb{X}$ is a smooth transformation from an open set $\mathbb{X}$ of $\mathbb{R}^n$ into $\mathbb{X}$, called *phase space*. The notation Φ^k stands for the k-iterate of the transformation Φ in (6.15), that is,

$$\Phi^k = \underbrace{\Phi \circ \cdots \circ \Phi}_{k \text{ times}} \ , \tag{6.16}$$

The family $(\Phi^k)_{k \in \mathbb{N}}$ of transformations of the phase space $\mathbb{X}$ is called *discrete-time flow*.

Remark 6.6 The discrete-time flow may be deduced by sampling, at period $\Delta T > 0$, from the continuous flow (2.39) associated with a continuous-time dynamical system (see Theorem 2.13), hence coinciding with $(\Phi_{k\Delta T})_{k \in \mathbb{N}}$. ◇

6.3.1 Stability of an Equilibrium Point

Equilibrium points are steady states of the classical dynamical system (6.15).

Definition 6.7 *An* equilibrium *point* x_E *of the classical dynamical system* (6.15) *is a fixed point of the transformation* Φ*, that is,*

$$\Phi(x_E) = x_E \ . \tag{6.17}$$

We now introduce two notions related to stability, counterparts of those introduced in § 4.2 for continuous-time classical dynamical systems. An equilibrium point x_E is *stable* if, when starting close enough to x_E, all the states x_k visited by the trajectory $(x_k)_{k \in \mathbb{N}}$ generated by (6.15) remain close to x_E. It is said to be *attractive* if the trajectory converges towards x_E. The attractive character of an equilibrium point is an *asymptotic* property, whereas stability concerns the whole trajectory for all times $k \in \mathbb{N}$, hence it is both a *transient* and *asymptotical* property. An equilibrium point is *asymptotically stable* if it is both stable and attractive.

Definition 6.8 *An equilibrium point x_E of the discrete-time classical dynamical system* (6.15) *is said to be*

- stable *if, for all neighborhood W' of x_E, there exists a neighborhood W'' of x_E such that, for all state x in W'', the k-iterate $\Phi^k(x)$ is defined and belongs to W' for all $k \in \mathbb{N}$ (an equilibrium is said to be* unstable *if it is not stable);*
- attractive *if there exists a neighborhood W' of x_E having the following property: for all state x in W', the trajectory $(\Phi^k(x))_{k\in\mathbb{N}}$ of k-iterates tends towards x_E when $k \to +\infty$;*
- asymptotically stable *if it is both stable and attractive.*

The notion of *Lyapunov function* is shoehorned to characterizing stability without having to know the trajectories.

Definition 6.9 *Let x_E be an equilibrium point of the classical dynamical system* (6.15). *We call* Lyapunov function *for the transformation Φ in a neighborhood of x_E a continuous function $\mathfrak{V}$, defined in a neighborhood of x_E such that:*

- $\mathfrak{V}(x) > 0$ *except at $x = x_E$ where $\mathfrak{V}(x_E) = 0$;*
- *the function $\mathfrak{V}$ satisfies the inequality*

$$\dot{\mathfrak{V}}(x) := \mathfrak{V}\big(\Phi(x)\big) - \mathfrak{V}(x) \le 0 \,. \tag{6.18}$$

Lyapunov functions are helpful to prove stability as follows.

Proposition 6.10 *If there exists a Lyapunov function $\mathfrak{V}$ for the transformation Φ in a neighborhood of the equilibrium x_E, then x_E is a stable equilibrium point of the classical dynamical system* (6.15). *In particular, for any state x in a neighborhood of x_E, the k-iterates $\Phi^k(x)$ are defined for all $k \in \mathbb{N}$.*

Moreover, assume that the function $\mathfrak{V}$ is defined in a neighborhood W of x_E, and that the set $\{x \in W \mid \dot{\mathfrak{V}}(x) = 0\}$ does not contain any subset invariant by the transformation Φ other than the singleton $\{x_E\}$. Then x_E is an asymptotically stable equilibrium of (6.15).

Proof The proof is inspired by [28].

Let W be a neighborhood of x_E on which $\mathfrak{V}$ is a Lyapunov function for the transformation Φ. Consider a compact neighborhood K of x_E included in W. For $\varepsilon > 0$, consider $V_\varepsilon = \{x \in K \mid \mathfrak{V}(x) \le \varepsilon\} \subset K \subset W$. We show, as in Proposition 4.30, that the family $(V_\varepsilon)_{\varepsilon>0}$ forms a fundamental system of neighborhoods of x_E. This makes it possible to work with these neighborhoods in the proofs of stability and of asymptotic stability.

For $\varepsilon > 0$ and $x \in V_\varepsilon$, we have that $x \in W$. Thus, by (6.18), we have that $\mathfrak{V}(\Phi(x)) \le \mathfrak{V}(x) < \varepsilon$, that is, $\Phi(x) \in V_\varepsilon \subset W$. We deduce that $\Phi^k(x)$ is defined for all $k \in \mathbb{N}$. Moreover, by (6.18), we have the inequalities

$$0 \le \mathfrak{V}(\Phi^k(x)) \le \mathfrak{V}(x) \,, \quad \forall k \in \mathbb{N} \,, \tag{6.19}$$

and it is straightforward that, for all $\varepsilon > 0$, one has that

$$x \in V_\varepsilon \Rightarrow \Phi^k(x) \in V_\varepsilon , \quad \forall k \in \mathbb{N} . \tag{6.20}$$

By (6.20) and the property that the family $(V_\varepsilon)_{\varepsilon>0}$ forms a fundamental system of neighborhoods of x_{E}, we conclude that the equilibrium point x_{E} is stable.

Now, suppose that the set $\{x \in W \mid \dot{\mathfrak{V}}(x) = 0\}$ does not contain any subset invariant by the transformation Φ other than the singleton $\{x_{\mathrm{E}}\}$, and that the equilibrium x_{E} is not attractive. Then, there exists an $\varepsilon_0 > 0$ and a point $x_0 \in V_{\varepsilon_0}$ such that $\Phi^k(x_0) \not\to x_{\mathrm{E}}$. That is, there exists $\eta > 0$ and a sequence $(k_m)_{m\in\mathbb{N}}$ increasing towards $+\infty$ such that:

$$\Phi^{k_m}(x_0) \notin V_\eta . \tag{6.21}$$

Now, by (6.19), the sequence of points $\big(\Phi^{k_m}(x_0)\big)_{m\in\mathbb{N}}$ belongs to the bounded set V_{ε_0}, hence admits a subsequence (still denoted $\big(\Phi^{k_m}(x_0)\big)_{m\in\mathbb{N}}$) which converges towards a point $\overline{x}$. We now show that $\overline{x} = x_{\mathrm{E}}$, which contradicts (6.21) and our assumption that the equilibrium x_{E} is not attractive. Indeed, the sequence of general term $\mathfrak{V}\big(\Phi^k(x_0)\big)$ is decreasing and positive, hence has a limit $l \geq 0$. Moreover, for any integer j, the sequences $\mathfrak{V}\big(\Phi^{k_m}(x_0)\big)_{m\in\mathbb{N}}$ and $\mathfrak{V}\big(\Phi^{j+k_m}(x_0)\big)_{m\in\mathbb{N}}$ are two subsequences of $(\mathfrak{V}\big(\Phi^k(x_0)\big))_{k\in\mathbb{N}}$ which satisfy:

$$\begin{aligned} l &= \lim_{m\to+\infty} \mathfrak{V}\big(\Phi^{k_m}(x_0)\big) &&= \mathfrak{V}(\overline{x}) \\ l &= \lim_{m\to+\infty} \mathfrak{V}\big(\Phi^{j+k_m}(x_0)\big) &&= \mathfrak{V}\big(\Phi^j(\overline{x})\big) . \end{aligned}$$

Therefore, the scalar $\mathfrak{V}\big(\Phi^j(\overline{x})\big)$ is independent of j, which implies that $\dot{\mathfrak{V}}\big(\Phi^j(\overline{x})\big) = 0$. As a result, all the sequence $\big(\Phi^j(\overline{x})\big)_{j\in\mathbb{N}}$ is contained in the set $\{x \in W \mid \dot{\mathfrak{V}}(x) = 0\}$. Since the orbit $\{\Phi^j(\overline{x}), j \in \mathbb{N}\}$ is clearly invariant by the transformation Φ, it is, by assumption, necessarily equal to the singleton $\{x_{\mathrm{E}}\}$. Thus $\overline{x} = \Phi^j(\overline{x}) = x_{\mathrm{E}}$. This contradicts (6.21), and this ends the proof. □

6.3.2 Case of Discrete-Time Linear Dynamical Systems

We associate with an $n \times n$ matrix F the *discrete-time linear classical dynamical system*

$$x_{k+1} = F x_k , \quad k \in \mathbb{N} . \tag{6.22}$$

In other words, the phase space is $\mathbb{R}^n$, the transformation Φ in (6.15) is the linear mapping $x \mapsto Fx$, and the discrete-time flow in (6.16) is given by

$$x \mapsto F^k x , \quad k \in \mathbb{N} . \tag{6.23}$$

We now discuss how the asymptotic behavior of the trajectories of the linear dynamical system (6.22) is related to the modulus of the eigenvalues of the matrix F. Recall that $\mathcal{S}(F)$ denotes the set of eigenvalues of F (see Definition 2.24).

Proposition 2.26 makes it possible to obtain an expression of the flow (6.23) in projection on the *complex* decomposition (2.52) of the space $\mathbb{C}^n$.

Proposition 6.11 *For all $x \in \mathbb{R}^n$ and k integer large enough, we have that:*

$$F^k x = \sum_{\lambda \in \mathcal{S}(F)} \sum_{l=0}^{\nu(\lambda)-1} C_k^l \lambda^{k-l} (F - \lambda I)^l p_\lambda(x) \,. \tag{6.24}$$

Proof By linearity, it is enough to compute

$$F^k p_\lambda(x) = (F - \lambda I + \lambda I)^k p_\lambda(x) = \sum_{l=0}^{\nu(\lambda)-1} C_k^l \lambda^{k-l} (F - \lambda I)^l p_\lambda(x) \,,$$

as soon as $k \geq \nu(\lambda)$. □

We now discuss the stability properties of the equilibrium point $x_{\mathrm{E}} = 0$ of the linear dynamical system (6.22) in function of the eigenvalues of the square matrix F.

Proposition 6.12 *Let $\lambda_1, \ldots, \lambda_r$ denote the distinct eigenvalues of the square matrix F. Consider the equilibrium point $x_{\mathrm{E}} = 0 \in \mathbb{R}^n$ of the linear dynamical system* (6.22).

1. *If there exists at least one $i = 1, \ldots, r$ such that $|\lambda_i| > 1$, then the equilibrium point 0 is unstable.*
2. *If for all $j = 1, \ldots, r$, one has that $|\lambda_j| \leq 1$, then*
   (a) *if $\forall j = 1, \ldots, r$, $|\lambda_j| < 1$, then the equilibrium point 0 is asymptotically stable;*
   (b) *if $\exists i = 1, \ldots, r$, $|\lambda_i| = 1$ and $\nu(\lambda_i) > 1$, then the equilibrium point 0 is unstable;*
   (c) *if $\exists i = 1, \ldots, r$, $|\lambda_i| = 1$ and if $\forall j = 1, \ldots, r$, $(|\lambda_j| = 1 \Rightarrow \nu(\lambda_j) = 1)$, then the equilibrium point 0 is stable but not asymptotically stable.*

Proof The proof boils down to comparing an exponential growth $k^{\nu(\lambda_j)-1}$ to a polynomial growth λ^k in the asymptotical formula (6.24) when k goes to $+\infty$.

1. Since $\mathrm{Ker}(F - \lambda_j I)^{\nu(\lambda_j)} \backslash \mathrm{Ker}(F - \lambda_j I)^{\nu(\lambda_j)-1}$ is not reduced to the singleton $\{0\}$, by definition of the eigenvalue index $\nu(\lambda_j)$ (see Definition 2.24), we can pick up a nonzero complex vector $z \in \mathcal{N}(\lambda_j) \backslash \mathrm{Ker}(F - \lambda_j I)^{\nu(\lambda_j)-1}$. From z, we build a nonzero real vector $x = z + \overline{z}$ by using the conjugate mapping recalled in § 2.1. Then, we observe in (6.24) that $\|F^k x\|$ grows exponentially towards $+\infty$ (at rate at least λ_j^k): the equilibrium point 0 is thus unstable.

2. (a) Let λ be such that $\max\{|\lambda_1|, \ldots, |\lambda_r|\} < \lambda < 1$. We deduce from (6.24) that there exists a constant M such that

$$\|F^k x\| \leq M\lambda^k \|x\| \ , \ \forall x \in \mathbb{R}^n \ . \tag{6.25}$$

We conclude that the equilibrium point 0 is asymptotically stable.

(b) We pick up x as in item 1. We observe in (6.24) that $k \mapsto \|F^k x\|$ grows as a polynomial $k^{\nu(\lambda_j)-1}$, where $\nu(\lambda_i) - 1 \geq 1$, towards $+\infty$: the equilibrium point 0 is thus unstable.

(c) We observe in (6.24) that there exists a constant M such that $\|F^k x\| \leq M\|x\|$, for all $x \in \mathbb{R}^n$: the equilibrium point 0 is thus stable. Then, we choose x as in item 1 and observe that $F^k x$ does no tend to zero when k goes to $+\infty$. Therefore, the equilibrium point 0 is not asymptotically stable.

This concludes the proof. □

Thanks to the above result, we now have a simple criterion of asymptotic stability of the equilibrium point $x_{\mathrm{E}} = 0$ of the linear dynamical system (6.22).

Theorem 6.13 *The equilibrium point* 0 *of the linear dynamical system* (6.22) *is asymptotically stable if, and only if, all the eigenvalues of the matrix F have modulus strictly less than one.*

Definition 6.14 *We call* stability disk *the set* $\{s \in \mathbb{C} \mid |s| < 1\}$ *of complex numbers having modulus strictly less than one. We say that a square matrix is* asymptotically stable *if all its eigenvalues belong to the stability disk.*

Remark 6.15 We also use the terminology *stable matrix* and *unstable matrix* in the corresponding cases of Proposition 6.12. ◇

As in the continuous-time case, there exists a so-called *Jury criterion*, which makes it possible to check the modulus of the roots of the characteric polynomial of a square matrix, hence to test asymptotic stability without having to compute the eigenvalues.

Lyapunov functions can be explicitely given in the linear case. This can prove useful for the study of nonlinear dynamical systems in the neighborhood of an equilibrium (see the proof of Proposition 6.17).

Proposition 6.16 *The equilibrium point $x_{\mathrm{E}} = 0$ of the linear dynamical system* (6.22) *is asymptotically stable if, and only if, for all positive definite matrix Q, there exists a positive definite matrix P such that:*

$$F^{\top} P F - P = -Q \ . \tag{6.26}$$

In that case, the function $\mathfrak{V}(x) = x^{\top} P x$ is a Lyapunov function for the equilibrium $x_{\mathrm{E}} = 0$ of (6.22).

Proof The condition (6.26) is sufficient for $\mathfrak{V}$ to be a Lyapunov function for the equilibrium $x_{\mathrm{E}} = 0$ of (6.22). Indeed, consider $Q = I$, and let P be a positive definite matrix such that $F^\top P F - F = -I$. Then, $\mathfrak{V}(x) = x^\top P x$ is a Lyapunov function for (6.22). Indeed, it satisfies

$$\dot{\mathfrak{V}}(x) = x^\top F^\top P F x - x^\top P x = -x^\top x \leq 0 ,$$

where $\dot{\mathfrak{V}}(x)$ is defined in (6.18). As a consequence, $\{x \in \mathbb{R}^n \mid \dot{\mathfrak{V}}(x) = 0\} = \{0\}$. Hence, by Proposition 6.10, the equilibrium point $x_{\mathrm{E}} = 0$ of (6.22) is asymptotically stable.

On the other hand, suppose that the equilibrium point $x_{\mathrm{E}} = 0$ of (6.22) is asymptotically stable. By Theorem 6.13, the matrix F has all its eigenvalues with modulus strictly less than one. This implies, by (6.24), that the following series converges

$$P = \sum_{i=0}^{+\infty} (F^\top)^i Q F^i .$$

Moreover, the limit P is a positive definite matrix which satisfies

$$F^\top P F - P = \sum_{i=1}^{+\infty} (F^\top)^i Q F^i - \sum_{i=0}^{+\infty} (F^\top)^i Q F^i = -Q .$$

This concludes the proof. □

A discrete-time counterpart of the continuous-time perturbation Theorem 4.23 is the following proposition, which makes use of the Jacobian matrix recalled in Remark 4.19.

Proposition 6.17 *If x_{E} is an equilibrium point of the classical dynamical system* (6.15) *such that the Jacobian matrix $\frac{\partial \Phi}{\partial x}(x_{\mathrm{E}})$ is asymptotically stable, then x_{E} is an asymptotically stable equilibrium point of the system* (6.15).

Proof By Proposition 6.10, it is enough to display a Lyapunov function $\mathfrak{V}$ for Φ in a neighborhood W of x_{E} such that the set $\{x \in W \mid \dot{\mathfrak{V}}(x) = 0\}$ is reduced to the singleton $\{x_{\mathrm{E}}\}$. Set $F = \frac{\partial \Phi}{\partial x}(x_{\mathrm{E}})$. Let P be a symmetric positive definite matrix such that $F^\top P F - P = -I$ (the existence is assured by Proposition 6.16). We now show that the function $\mathfrak{V}(x) = (x - x_{\mathrm{E}})^\top P (x - x_{\mathrm{E}})$ answers the question. By definition of the Jacobian matrix F, we can write

$$\Phi(x) = x_{\mathrm{E}} + F(x - x_{\mathrm{E}}) + \varepsilon(x - x_{\mathrm{E}})$$

where $\lim_{\|x\|\to 0} \varepsilon(x) = 0$. Then, we have that

$$\begin{aligned}
\dot{\mathfrak{V}}(x) &= \mathfrak{V}\big(\Phi(x)\big) - \mathfrak{V}(x) \\
&= \mathfrak{V}\big(x_{\mathrm{E}} + F(x - x_{\mathrm{E}}) + \varepsilon(x - x_{\mathrm{E}})\big) - \mathfrak{V}(x) \\
&= \big(F(x - x_{\mathrm{E}}) + \varepsilon(x - x_{\mathrm{E}})\big)^{\top} P(F(x - x_{\mathrm{E}}) + \varepsilon(x - x_{\mathrm{E}})) \\
&\quad -(x - x_{\mathrm{E}})^{\top} P(x - x_{\mathrm{E}}) \\
&= -\|x - x_{\mathrm{E}}\|^2 + (x - x_{\mathrm{E}})^{\top} F^{\top} P\varepsilon(x - x_{\mathrm{E}}) \\
&\quad +\varepsilon(x - x_{\mathrm{E}})^{\top} PF(x - x_{\mathrm{E}}) + \varepsilon(x - x_{\mathrm{E}})^{\top} P\varepsilon(x - x_{\mathrm{E}}) \\
&\leq -(1 - \eta)\|x - x_{\mathrm{E}}\|^2 ,
\end{aligned}$$

for at least one $\eta \in]0, 1[$ and for x in a sufficiently small neighborhood of x_{E}. As a consequence, the function $\mathfrak{V}$ is a Lyapunov function for Φ in a neighborhood of x_{E}, and it is such that the set $\{x \in W \mid \dot{\mathfrak{V}}(x) = 0\}$ is reduced to the singleton $\{x_{\mathrm{E}}\}$.

6.4 Stability of Controlled Discrete-Time Linear Dynamical Systems

The links between stability for continuous-time and discrete-time systems are strong. We now shed light onto the BIBS-stability of discretized systems of the form (6.3).

Definition 6.18 *The linear dynamical system* (6.3) *is said to be* BIBS-stable *if, for all initial state x_0 and for all bounded input $(u_k, k \in \mathbb{N})$, the state trajectory $(x_k, k \in \mathbb{N})$ generated by* (6.3) *is bounded, that is, if*

$$\sup_{k\in\mathbb{N}} \|u_k\| < +\infty \Rightarrow \sup_{k\in\mathbb{N}} \|x_k\| < +\infty . \tag{6.27}$$

The linear dynamical system (6.3) *is said to be* BIBO-stable *if, for all initial state x_0 and for all bounded input $(u_k, k \in \mathbb{N})$, the output trajectory $(y_k, k \in \mathbb{N})$ generated by* (6.3) *is bounded, that is, if*

$$\sup_{k\in\mathbb{N}} \|u_k\| < +\infty \Rightarrow \sup_{k\in\mathbb{N}} \|y_k\| < +\infty . \tag{6.28}$$

The following proposition is the discrete-time counterpart of the continuous-time Proposition 5.8.

Proposition 6.19 *If all the eigenvalues of the matrix A_d belong to the stability disk, then the linear dynamical system* (6.3) *is* BIBS*-stable. If A_d has at least an eigenvalue with modulus strictly larger than one, then the linear dynamical system* (6.3) *is not* BIBS*-stable.*

Proof By Proposition 6.11, for all $x \in \mathbb{R}^n$ and k large enough, we have (with the notations of § 6.3.2):

$$A_d^k x = \sum_{\lambda \in \mathcal{S}(A_d)} \sum_{i=0}^{\nu(\lambda)-1} C_k^i \lambda^{k-i} (A_d - \lambda I)^i p_\lambda(x) . \tag{6.29}$$

If all the eigenvalues of A_d have modulus strictly less than one, then it follows from (6.7) and (6.29) that a bounded control produces a bounded state trajectory.

If A_d has at least an eigenvalue with modulus strictly larger than one, select any initial state x_0 having nonzero projection on the eigenspace associated with this eigenvalue (see the proof of Proposition 6.12). Then, the zero control $u_k \equiv 0$ yields a state trajectory $(x_k, k \in \mathbb{N})$ which is not bounded. □

REMARKS

- Notice that this proposition does not state a necessary and sufficient condition of BIBS-stability. Indeed, we refer the reader to Proposition 6.12, where we have discussed the many cases associated with having at least one eigenvalue with modulus 1.
- The parallel between continuous-time and discrete-time linear dynamical systems stability may be stressed by recalling that, by (6.4), A_d is nothing but the matrix $e^{\Delta T A}$. As every complex matrix is equivalent to a triangular matrix on $\mathbb{C}$, there exists a complex matrix P such that $A = PSP^{-1}$, where the diagonal of S carries the eigenvalues of A, and $e^{A \Delta T} = Pe^{\Delta T S}P^{-1}$, where the diagonal of $e^{\Delta T S}$ carries the eigenvalues of $e^{\Delta T A}$. Thus, if the eigenvalues of A all have strictly negative real part, the eigenvalues of A_d belong to the interior of the unit circle.

6.5 Controllability. Regulator

Now, as we did for continuous-time linear dynamical systems in § 5.4, we highlight controllability issues in discrete-time.

Definition 6.20 *We say that the linear dynamical system* (6.3) *is* controllable *if, for all couple of vectors* (x_i, x_f) *of* $\mathbb{R}^n$*, there exist a positive integer* $\bar{k}$ *and a control sequence* $(u_k, k = 0, \dots, \bar{k})$ *that, when driving the system* (6.3) *with initial state* $x_0 = x_i$*, yield* $x_{\bar{k}} = x_f$.

There is a simple algebraic characterization of controllability, the *Kalman criterion of controllability*.

Theorem 6.21 *The linear dynamical system* (6.3) *is controllable if, and only if, the* controllability matrix

$$\mathcal{C}_d := \begin{pmatrix} B_d & A_d B_d & \cdots & A_d^{n-1} B_d \end{pmatrix} \tag{6.30}$$

has rank n. We also say that the couple (A_d, B_d) *is* controllable.

Proof Suppose that the linear dynamical system (6.3) is controllable. Let us pick up $(x_i, x_f) = (0, x_a)$, so that $x_0 = 0$. By (6.7), we have that

$$x_k = \sum_{i=0}^{k-1} A_d^{k-1-i} B_d u_i \ .$$

To reach any x_a in $\mathbb{R}^n$, it is necessary that, for k large enough:

$$\mathbb{R}^n = \operatorname{Im}(B_d, A_d B_d, \cdots, A_d^{k-1} B_d) \ .$$

By the Cayley-Hamilton theorem 4.10, we have the following equality, for all $k \geq n$:

$$\operatorname{Im}(B_d, A_d B_d, \cdots, A_d^{k-1} B_d) = \operatorname{Im}(B_d, A_d B_d, \cdots, A_d^{n-1} B_d) \ .$$

Thus, it is necessary that the rank of $\mathcal{C}_d$ be equal to n for the linear dynamical system (6.3) to be controllable.

Consider a given couple (x_i, x_f), and suppose that $\mathcal{C}_d$ has rank n. Therefore, there exists a sequence of controls $u_0, \ldots, u_n$ such that:

$$x_f - A_d^n x_i = (B_d, A_d B_d, \cdots, A_d^{n-1} B_d) \begin{pmatrix} u_{n-1} \\ u_{n-2} \\ \vdots \\ u_0 \end{pmatrix} .$$

Thus, for this sequence of controls, when $x_0 = x_d$, we reach $x_n = x_f$ by (6.7). □

The proof of the following corollary (the counterpart of Corollary 5.17) is easy.

Corollary 6.22 *The following conditions are equivalent.*

1. *The linear dynamical system* (6.3) *is controllable.*
2. *The couple* (A_d, B_d) *is controllable.*
3. *Any vector* x *such that* $x^\top B_d = x^\top A_d B_d = \cdots = x^\top A_d^{n-1} B_d = 0$ *is zero.*
4. *The symmetric positive matrix*

$$P_c = \sum_{j=0}^{n-1} A_d^j B_d B_d^\top (A_d^\top)^j \tag{6.31}$$

is definite (or invertible).

In the case where the linear dynamical system (6.3) is not controllable, one can decompose the state-space $\mathbb{R}^n$ in controllable and uncontrollable parts, as for continuous-time linear dynamical systems.

If the linear dynamical system (6.3) is controllable, we can define the controllable companion form and, as for continuous-time linear systems, we show the *regulator modes placement* Theorem. The proof is similar to that of Theorem 5.27.

Theorem 6.23 *Under the assumption of controllability of the linear dynamical system* (6.3)*, there exists a* linear state feedback

$$u_k = -Kx_k \ , \tag{6.32}$$

where K is the gain matrix *of counter-reaction, such that the state matrix $(A_d - B_d K)$ of the* closed-loop *linear dynamical system* (6.3)–(6.32) *is asymptotically stable.*

6.6 Observability. Observer

When the state is only partially observed, one can, as discussed in § 5.5 in continuous-time, reconstitute the whole state under the assumption of observability.

Definition 6.24 *The linear dynamical system* (6.3) *is* observable *if, for all vector x_i of $\mathbb{R}^n$, there exists a time $\bar{k} \in \mathbb{N}$ and a sequence of controls $(u_l, l = 0, \ldots, \bar{k})$ such that the initial state $x_0 = x_i$ may be determined from the sequence of outputs $(y_l, l = 0, \ldots, \bar{k})$ given by* (6.3).

As for controllability, we have the *Kalman observability criterion.*

Theorem 6.25 *The system* (6.3) *is observable if, and only if, the* observability matrix

$$\mathcal{O}_d = \begin{pmatrix} C \\ CA_d \\ \vdots \\ CA_d^{n-1} \end{pmatrix} \tag{6.33}$$

has rank n. We also say that the couple (A_d, C) *is* observable.

Proof The output of the system (6.3) satisfies, for all integer k:

$$y_k = CA_d^k x_0 + \sum_{i=0}^{k-1} CA_d^{k-1-i} B_d u_i \ . \tag{6.34}$$

Thus, we have that

$$\begin{pmatrix} C \\ CA_d \\ \vdots \\ CA_d^k \end{pmatrix} x_0 = \begin{pmatrix} y_0 \\ y_1 \\ \vdots \\ y_k \end{pmatrix} - \begin{pmatrix} 0 & . & . & . & 0 \\ CB_d & 0 & . & . & 0 \\ . & . & . & . & . \\ . & . & . & . & . \\ CA_d^{k-1}B_d & . & . & CB_d & 0 \end{pmatrix} \begin{pmatrix} u_0 \\ u_1 \\ \vdots \\ u_k \end{pmatrix} .$$

The initial state x_0 is obtained from $y_0, \ldots, y_k$ and $u_0, \ldots, u_k$ if, and only if, the linear mapping defined by $\left(C, CA_d, \ldots, CA_d^k\right)^\top$ is left-invertible, for given k, thus if, and only if, the rank of $\left(C, CA_d, \ldots, CA_d^k\right)$ is equal to n. Now, by the Cayley-Hamilton theorem 4.10, if such an integer k exists, it can be taken equal to $(n-1)$. This ends the proof. □

The proof of the following corollary (the counterpart of Corollary 5.32) is easy.

Corollary 6.26 *The following conditions are equivalent.*

1. *The linear dynamical system* (6.3) *is observable.*
2. *The couple* (A_d, C) *is observable.*
3. *Any vector* x *such that* $Cx = CA_dx = \cdots = CA_d^{n-1}x = 0$ *is zero.*
4. *The symmetric positive matrix*

$$P_o = \sum_{j=0}^{n-1} (A_d^\top)^j C^\top C A_d^j \tag{6.35}$$

is definite (or invertible).

In the case where the linear dynamical system (6.3) is not observable, one can decompose the state-space $\mathbb{R}^n$ into observable and unobservable parts, as for continuous-time linear dynamical systems (see Proposition 5.34).

If the linear dynamical system (6.3) is observable, we can define the observable companion form and, as for continuous-time linear systems, we show the *observer modes placement* Theorem. The proof is similar to that of Theorem 5.39.

Theorem 6.27 *Under the assumption of observability of the linear dynamical system* (6.3)*, one can build a* linear asymptotic observer *of* (6.3)*, or* Luenberger observer, *with internal variable* $\hat{x}$ *and external variables* u *and* y*, as follows:*

$$\hat{x}_{k+1} = A_d\hat{x}_k + B_du_k - L(C\hat{x}_k - y_k) . \tag{6.36}$$

The gain matrix L *is chosen so that the matrix* $(A_d - LC)$ *of the dynamics of the error* $e_k := \hat{x}_k - x_k$ *of the closed system* (6.3)–(6.36) *be asymptotically stable.*

In the expression (6.36), notice that the state $\hat{x}_{k+1}$ of the observer at time $k+1$ depends upon the measurements $y_0, \ldots, y_k$ up to time k. We now display an asymptotic observer, the state $\hat{x}_{k+1}$ of which at time $k+1$ depends upon the measurements $y_0,\ldots,$ y_k, y_{k+1} up to time $k+1$. Therefore, the last measurement makes it possible to improve the knowledge of the state at time $k+1$.

Theorem 6.28 *Under the assumption of observability of the linear dynamical system* (6.3), *one can build an asymptotic observer of* (6.3), *with internal variable $\hat{x}$ and external variables u and y, as follows:*

$$\hat{x}_{k+1} = A_d\hat{x}_k + (B_d - LCB_d)u_k - L(CA_d\hat{x}_k - y_{k+1}) \,. \tag{6.37}$$

The gain matrix L is chosen so that the matrix $(A_d - LC)$ of the dynamics of the error $e_k = \hat{x}_k - x_k$ of the closed system (6.3)–(6.37) *be asymptotically stable.*

We discuss in Chap. 7 the links between this last observer and the Kalman-Bucy filter.

6.7 Observer-Regulator Synthesis. Separation Principle

As for controllable and observable linear dynamical systems in continuous-time, those in discrete-time satisfy the *estimation-control separation principle*, yet discussed in § 5.6. On the one hand, a regulator is designed. On the other hand, an observer is built. Then, though designed separately, their combination leads to stabilization of the closed system.

Definition 6.29 *We call* observer-regulator *of the linear dynamical system* (6.3) *a linear dynamical system with input y, with state $\hat{x}$ and with output u of the form*

$$\begin{cases} \hat{x}_{k+1} = A_d\hat{x}_k + B_d u_k - L(C\hat{x}_k - y_k) \\ u_k = -K\hat{x}_k \,, \end{cases} \tag{6.38}$$

as illustrated on Fig. 5.1.

Remark 6.30 The observer-regulator (6.38) corresponds to Theorem 6.27. Another possible form is given below in (6.40), corresponding to Theorem 6.28. ◇

The following theorem expresses the so-called *estimation-control separation principle.*

Theorem 6.31 *Suppose that the linear dynamical system* (6.3) *is controllable and observable, and let K and L be gain matrices such that the matrices $(A_d - B_dK)$ and $(A_d - LC)$ are asymptotically stable. Then, the linear dynamical system* (6.3) *closed with the observer-regulator* (6.38) *forms a* closed-loop system *with state $(x, e)^\top = (x, \hat{x} - x)^\top$ and dynamics given by*

$$\begin{cases} x_{k+1} = (A_d - B_d K)x_k - B_d K e_k \\ e_{k+1} = (A_d - LC)e_k \ , \end{cases} \tag{6.39}$$

such that the zero equilibrium $(x, e)^\top = (0, 0)^\top$ *is asymptotically stable (for all initial condition* x_0 *of the state and* $\hat{x}_0$ *of the observer).*

Proof Using the relations $y_k = Cx_k$ in (6.3) and $u_k = -K\hat{x}_k$ in (6.38), the linear dynamical system (6.3)–(6.38) writes:

$$\begin{cases} x_{k+1} = A_d x_k - B_d K \hat{x}_k \\ \hat{x}_{k+1} = (A_d - LC - B_d K)\hat{x}_k + LCx_k \ . \end{cases}$$

Defining the estimation error vector $e_k = \hat{x}_k - x_k$, we obtain the following classical linear dynamical system (closed-loop):

$$\begin{pmatrix} x_{k+1} \\ e_{k+1} \end{pmatrix} = \begin{pmatrix} A_d - B_d K & -B_d K \\ 0 & A_d - LC \end{pmatrix} \begin{pmatrix} x_k \\ e_k \end{pmatrix} .$$

Since the hereabove matrix is block-diagonal, the closed-loop modes of the above system are given by the union of the modes of the regulator (eigenvalues of $A_d - B_d K$) and of the observer (eigenvalues of $A_d - LC$). The system is thus stabilized in closed-loop as soon as the gains K and L are chosen in such a way that the matrices $(A_d - B_d K)$ and $(A_d - LC)$ have their eigenvalues in the stability disk (see Theorem 6.13 and Definition 6.28).

The estimation-control separation principle corresponding to the observer given by Theorem 6.28 is expressed by the following theorem.

Theorem 6.32 *Suppose that the linear dynamical system* (6.3) *is controllable and observable, and let* K *and* L *be gain matrices such that the matrices* $(A_d - B_d K)$ *and* $(A_d - LCA_d)$ *are asymptotically stable. Then, the linear dynamical system* (6.3) *closed with the observer-regulator*

$$\begin{cases} \hat{x}_{k+1} = A_d \hat{x}_k + (B_d - LCB_d)u_k - L(CA_d\hat{x}_k - y_{k+1}) \\ u_k = -K\hat{x}_k \end{cases} \tag{6.40}$$

forms a closed-loop system *with state* $(x, e)^\top = (x, \hat{x} - x)^\top$ *and dynamics given by*

$$\begin{cases} x_{k+1} = (A_d - B_d K)x_k - B_d K e_k \\ e_{k+1} = (A_d - LCA_d)e_k \ , \end{cases} \tag{6.41}$$

such that the zero equilibrium $(x, e)^\top = (0, 0)^\top$ *is asymptotically stable (for all initial condition* x_0 *of the state and* $\hat{x}_0$ *of the observer).*

6.8 Choice of the Sampling Period

When a controllable and observable linear dynamical system in continuous-time having conjugated complex modes is sampled, the discretized system can lose the property of controllability or of observability. More precisely, we have the following theorem (see for example [2]).

Theorem 6.33 *A controllable and observable linear dynamical system in continuous-time remains controllable and observable after discretization by a* ZOH *at frequency* $1/\Delta T$ *if, and only if, for each pair of distinct eigenvalues* $\{\lambda_i, \lambda_j\}$ $(i \neq j)$ *of the state matrix of the continuous-time system, we have that*

$$\Re\lambda_i = \Re\lambda_j \Rightarrow \Im(\lambda_i - \lambda_j) \neq \frac{2n\pi}{\Delta T}\ , \quad \forall n \in \mathbb{Z}^*\ . \tag{6.42}$$

Example 6.34 One easily shows that the continuous-time linear dynamical system (6.10) is controllable and observable by computing the controllability and observability matrices which both have rank 2. Now, let us compute the controllability and observability matrices of the exact discretized of (6.10), given by (6.11). By (6.30), we obtain

$$\mathcal{C}_d = \begin{pmatrix} B_d & A_d B_d \end{pmatrix} = \begin{pmatrix} 1 - \cos\Delta T & \cos\Delta T - \cos 2\Delta T \\ \sin\Delta T & -\sin\Delta T + 2\cos\Delta T \sin\Delta T \end{pmatrix}$$

and, by (6.33),

$$\mathcal{O}_d = \begin{pmatrix} C \\ CA_d \end{pmatrix} = \begin{pmatrix} 1 & 0 \\ \cos\Delta T & \sin\Delta T \end{pmatrix} .$$

These two matrices are of full rank if, and only if:

$$\Delta T \neq n\pi\ , \quad \forall n \in \mathbb{Z}^*\ .$$

Now, the eigenvalues of the matrix $\begin{pmatrix} 0 & 1 \\ -1 & 0 \end{pmatrix}$ in (6.10) are $\{-i, i\}$, so that the above condition is identical to that of the relation (6.42), which indeed writes:

$$\Im(\lambda_1 - \lambda_2) = \Im\big(i - (-i)\big) = 2 \neq \frac{2n\pi}{\Delta T}\ .$$

△

Example 6.35 In the case of the inverted pendulum on a cart, discussed at Examples 4.37 and 6.4, one can check on the expression of the eigenvalues in (4.39) that, for all $\Delta T > 0$, the discretized system is controllable and observable. The direct study of the rank of (6.30) with the matrices (6.13) would have been more difficult. △

6.9 Links with the Input-Output Representation

Now, we turn the spotlight onto the relations between state (or internal) representation and input-output (or external) representation introduced in Chap. 3, in the case of discrete-time l.c.s. systems. This is the discrete-time counterpart of the discussion in continuous-time in § 5.7.

6.9.1 Impulse Response, Transfer Matrix and Realization

The notion of unit impulse function introduced in Definition 3.7 and of impulse response introduced in Definition 3.9 for continuous-time l.c.s. systems are adapted to discrete-time as follows.

Definition 6.36 *The* unit impulse sequence *is the sequence* $(\delta_{0,k})_{k\in\mathbb{N}}$*, where* $\delta_{i,j}$ *denotes the* Kronecker symbol *(*$\delta_{i,j} = 0$ *if* $i \neq j$*, 1 else). The* impulse response *of an l.c.s. discrete-time system is the response to the unit impulse sequence.*

The output at time k of the linear dynamical system (6.1), expressed in state representation, is:

$$y_k = CA_d^k x_0 + \sum_{l=0}^{k-1} CA_d^{k-1-l} B_d u_l \ .$$

Considering that the initial state is zero ($x_0 = 0$) makes it possible to deduce the following proposition.

Proposition 6.37 *The* impulse response *of the linear dynamical system* (6.1)*, with zero initial state, is given by the sequence of matrices*

$$h_k = CA_d^{k-1} B_d \ \text{if } k \geq 1 \ , \ h_k = 0 \ \text{if } k \leq 0 \ . \tag{6.43}$$

We have the property of discrete-time convolution Z2 in § B.2:

$$y_k = (h \star u)_k = , \ \forall k \in \mathbb{N} \ . \tag{6.44}$$

Similarly to the continuous-time case, there exists a transformation which maps a convolution product into a usual product: the z-transform described in § B.2. The notion of transfer matrix introduced at Definition 3.14 for continuous-time l.c.s. systems is adapted to discrete-time as follows. The *transfer matrix* of a discrete-time l.c.s. system, with m-input vector u and p-output vector y, is the $p \times m$ matrix $H(z)$ such that

$$Y(z) = H(z)U(z) \ , \tag{6.45}$$

where $z \in \mathbb{C}$, and $U(z)$ (respectively $Y(z)$) denotes the z-transform of $(u_k)_{k\in\mathbb{N}}$ (if it exists), (respectively of $(y_k)_{k\in\mathbb{N}}$).

To obtain the transfer matrix of the linear dynamical system (6.1), we apply to Eq. (6.1) the properties of linearity Z1 and of shift Z3 in § B.2 of the z-transform, giving:

$$Y(z) = C(zI - A_d)^{-1} B_d U(z) . \tag{6.46}$$

One deduces straightforwardly the following Proposition.

Proposition 6.38 *The* transfer matrix *of the discrete-time linear dynamical system in state representation* (6.1) *is the $p \times m$ matrix given by:*

$$H(z) = C(zI - A_d)^{-1} B_d . \tag{6.47}$$

The transfer matrices being of the same form as in the continuous-time case, we can state the following proposition, the counterpart of Proposition 5.48.

Proposition 6.39 *The elements of the transfer matrix* (6.47) *are strictly proper rational functions in z.*

REMARKS

- In case there exists a direct link between the input and the output , that is, if

$$y_k = Cx_k + Du_k ,$$

  it is straightforward to check that the transfer matrix has the expression

$$H(z) = C(zI - A_d)^{-1} B_d + D , \tag{6.48}$$

  and the elements of the transfer matrix are rational functions whose numerator and denominators have the same degree.
- As discussed in Chap. 3, the notions of proper transfer and of causality of the corresponding linear dynamical system are intimately linked. This can be observed in the discrete-time case. Indeed, if the transfer matrix in discrete-time is of the form $H(z) = C(zI - A_d)^{-1} B_d + D$, the dynamical relations write:

$$\begin{cases} x_{k+1} = A_d x_k + B_d u_k \\ y_k = Cx_k + Du_k . \end{cases}$$

  If $D = 0$, the output at time k only depends on the input values at times strictly less than k and the system is thus *strictly causal*. If $D \neq 0$, the output at time k depends on the input values at times less than or equal to k: the system is only *causal* and not strictly causal.
- The relation (5.48) gives a systematic way to compute the state matrix A_d of the discrete-time system obtained by sampling at period ΔT a continuous-time system

with state matrix A. Indeed, we can write:

$$A_d = e^{A\Delta T} = \mathcal{L}^{-1}\left[(sI - A)^{-1}\right](\Delta T) \,. \tag{6.49}$$

By the definition of the impulse response and the property that the z-transform of a convolution product is a usual product, one deduces the following proposition.

Proposition 6.40 *The transfer matrix* (6.47) *is the z-transform of the impulse response* (6.43) *of the linear dynamical system given in discrete-time state representation* (6.3).

Similarly to the continuous-time realization issue discussed in § 5.7.2, when a transfer matrix $H(z)$ that represents a linear discrete-time system is given, one can look for the expression of corresponding state-models of the form (6.3), that is look for matrices A_d, B_d, C such that:

$$H(z) = C(zI - A_d)^{-1} B_d \,. \tag{6.50}$$

Definition 6.41 *Any triple $(A_d,\ B_d,\ C)$ of matrices satisfying* (6.50) *is called a* discrete-time state realization *of the transfer matrix $H(z)$.*

As in the continuous-time case (see Theorem 5.54), we have the following result.

Theorem 6.42 *A linear dynamical system in discrete-time state representation* (6.3) *is* minimal *if and only if it is controllable and observable.*

6.9.2 Stability and Poles. Jury Criterion

As in the continuous-time case, the BIBS stability discussed in § 6.4 is connected to the poles of the discrete-time linear dynamical system, that is, to the modes of the matrix A_d in (6.3).

Proposition 6.43 *If the poles are in the stability disk, the discrete-time linear dynamical system* (6.3) *is* BIBS*-stable.*

Remark 6.44 This is only a sufficient condition: a BIBO-stable or a BIBS-stable l.c.s. system can be realized with a state matrix A_d which is not asymptotically stable. This happens when the system has uncontrollable or unobservable modes. ◇

As in the continuous-time case (see Proposition 5.60), we have the following proposition.

Proposition 6.45 *The transfer function of the linear dynamical system* (6.3) *only depends on the controllable and observable modes.*

Remark 6.46 For discrete-time systems, the analogue to the Routh test is the Jury test, which makes it possible to check that a polynomial has all its roots in the stability disk. ◇

Considering a polynomial $P(z)$ with real coefficients

$$P(z) = a_0 z^n + a_1 z^{n-1} + a_2 z^{n-2} + a_3 z^{n-3} + \cdots + a_{n-1} z + a_n \ , \tag{6.51}$$

we build the following table:

$$\begin{array}{llll} a_0 & a_1 & \cdots & a_{n-1} \ a_n \\ a_n & a_{n-1} & \cdots & a_1 \ \ a_0 \\ a_0^{n-1} & a_1^{n-1} & \cdots & a_{n-1}^{n-1} \\ a_{n-1}^{n-1} & a_{n-2}^{n-1} & \cdots & a_0^{n-1} \\ \cdots & \cdots & \cdots & \\ a_0^0 \ . & & & \end{array} \tag{6.52}$$

The first row is made of the coefficients of the polynomial $P(z)$, and the second row is made of these same coefficients in reverse order. The other rows are obtained by induction as follows:

$$a_i^{k-1} = a_i^k - \alpha_k a_{k-i}^k \ \text{with} \ \alpha_k = \frac{a_k^k}{a_0^k} \ \text{and} \ \alpha_n = \frac{a_n}{a_0} \ . \tag{6.53}$$

The following theorem is known under the name of *Jury criterion* ([6, p. 97], [43]).

Theorem 6.47 *The polynomial $P(z)$ in* (6.51) *has all its roots with modulus strictly less than one if, and only if, all the a_0^k, $k = 0, \ldots, n-1$ in the table* (6.52) *given by* (6.53) *are not zero and have the same sign, the sign of a_0. When the a_0^k are not zero, the number of coefficients a_0^k, $k = 0, \ldots, n-1$ not having the sign of a_0 corresponds to the number of roots of $P(z)$ outside the stability disk.*

6.9.3 Zeros of a Discrete-Time Scalar l.c.s. System

As in the continuous-time case discussed in § 3.5, the zeros of the transfer function $H(z)$ of a discrete-time scalar l.c.s. system are related to the rejection of some class of inputs.

Consider a discrete-time scalar l.c.s. system with transfer function $H(z)$, excited by a sinusoidal input of pulsation ω:

$$u_k = e^{ik\omega \Delta T} \ , \ k \geq 0 \ , \ u_k = 0 \ , \ k < 0 \ . \tag{6.54}$$

The zeros of $H(e^{i\omega \Delta T})$ correspond to the frequencies asymptotically rejected by the discrete-time l.c.s. system having transfer function $H(z)$ (see Corollary 3.30).

As in the continuous-time context, it is also convenient to define the *unit-step sequence* as follows (see Definition 3.4).

Definition 6.48 *The* unit-step sequence *is the sequence defined by* $\mathfrak{E}_n = 1$ *if* $n \geq 0$ *and by* $\mathfrak{E}_n = 0$ *if* $n < 0$.

Applying the final value Theorem Z6 in § B.2 in the discrete-time case, we can prove the following proposition.

Proposition 6.49 *If the transfer function* $H(z)$ *has a zero at* $z = 1$*, step function inputs of the form* $b\ (\mathfrak{E}_k)_{k\in\mathbb{N}}$*,* $b \in \mathbb{R}$*, are asymptotically rejected by the discrete-time l.c.s. system having transfer function* $H(z)$*.*

This motivates the following definition (see Definition 3.33 for continuous-time systems).

Definition 6.50 *We call* static gain *of the discrete-time l.c.s. system with transfer function* $H(z)$ *the real number* $H(1)$*.*

Finally, we discuss the links between two input-output relations: that of an original linear dynamical system in continuous-time, and that of the corresponding exact discretized obtained by sampling at a period ΔT with a zero-order hold.

6.9.4 Relation Between an l.c.s. System in Continuous-Time and the Exact Discretized

We have established in § 6.2 the link between the state matrix A of a continuous-time linear dynamical system and the state matrix A_d of its exact discretized sampled to a period ΔT by a zero-order hold. As discussed in Remark 6.4, the poles ν_i of the transfer function of a discretized system are related to the poles λ_i of the transfer function of the continuous-time system by the formula $\nu_i = e^{\lambda_i \Delta T}$. Recall that the mapping $s \mapsto z = e^{s\Delta T}$ is a conformal transformation of $\mathbb{C}$ that maps the stability half-plane $\{s \in \mathbb{C} \mid \Re(s) < 0\}$ into the stability disk $\{z \in \mathbb{C} \mid |z| < 1\}$.

Regarding the zeros of a transfer function, the link is less simple than for poles. Before shedding light onto this issue, we first study the process of discretization by a zero-order hold (ZOH) from the input-output point of view.

Consider a continuous-time input trajectory u, and let us introduce the distribution (see the discussion on the Theory of Distributions in § 3.2.2):

$$u^\star(t) := \sum_{k=0}^{+\infty} u(k\Delta T)\delta_{k\Delta T}(t)\ . \tag{6.55}$$

This distribution corresponds to the sampled signal given by the sequence $\big(u(k\Delta T)\big)_{k\in\mathbb{N}}$. The action of a ZOH consists in maintaining constant on the sampling interval $[k\Delta T, (k+1)\Delta T[$ the signal u at the value $u(k\Delta T)$. The holded input

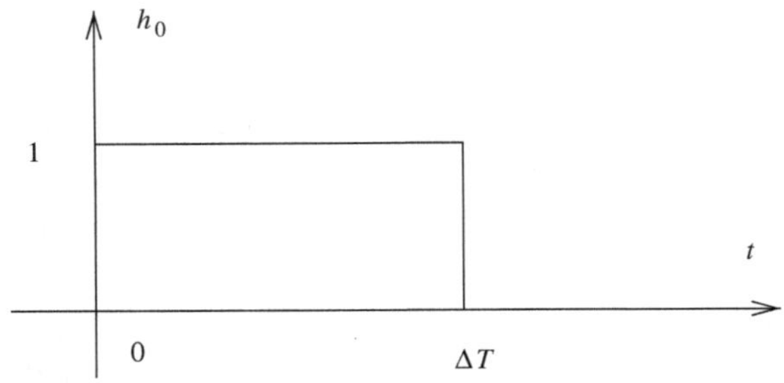

Fig. 6.2 The impulse response of a ZOH

signal $\overline{u}(t)$ obtained is given by the convolution of the sampled signal $u^\star(t)$ with the impulse response $h_0(t)$ of the ZOH represented in Fig. 6.2.

By definition of the Laplace transform (B.1), we easily show the following lemma.

Lemma 6.51 *The transfer function of a zero-order hold is given by:*

$$H_0(s) = \frac{1 - e^{-s\Delta T}}{s} . \tag{6.56}$$

Let us now consider a discrete-time l.c.s. system with transfer function $H(s)$ submitted to an input sampled and holded at period ΔT as seen on Fig. 6.3.

We prove the following theorem which links the transfer function of the discretized system to $P(s)$ given by the product:

$$P(s) = H(s)H_0(s) . \tag{6.57}$$

Theorem 6.52 *The transfer function $H_d(z)$ of the discretized system obtained by sampling and zero-order holding at period ΔT from a continuous-time l.c.s. system with transfer function $H(s)$ is given by*

$$H_d(z) = \sum_{k=0}^{+\infty} p(k\Delta T)z^{-k} , \tag{6.58}$$

where $p(t)$ is the inverse Laplace transform of $P(s)$ given by (6.57).

Proof By the property L2 of the Laplace transform in § B.1, if $\overline{U}(s)$ and $U^\star(s)$ denote the respective Laplace transforms of $\overline{u}(t)$ and $u^\star(t)$, if they exist, we have

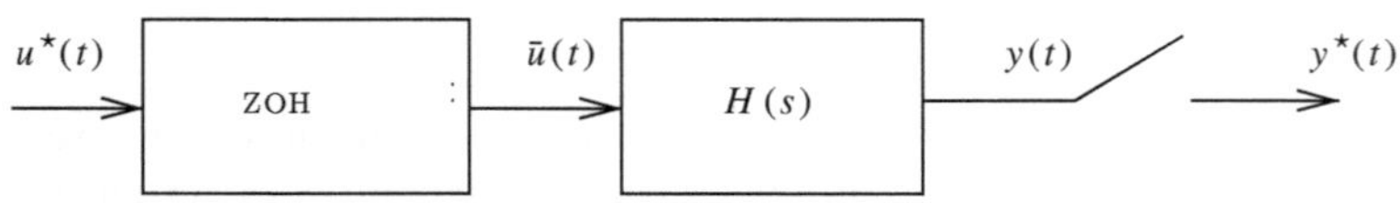

Fig. 6.3 Sampled and holded system

that

$$\overline{U}(s) = H_0(s)U^\star(s) ,$$

where, by (6.55) and Table B.1, $U^\star(s)$ is given by:

$$U^\star(s) = \sum_{k=0}^{+\infty} u(k\Delta T)e^{-k\Delta T s} .$$

The Laplace transform $Y(s)$ of the output $y(t)$ of the system can therefore be written as

$$Y(s) = H(s)\overline{u}(s) = H(s)H_0(s)U^\star(s) = \sum_{k=0}^{+\infty} P(s)u(k\Delta T)e^{-k\Delta T s} ,$$

where $P(s)$ is given by (6.57).

By the time shifting Theorem L3 in § B.1, we can write

$$y(t) = \mathcal{L}^{-1}\left[Y(s)\right](t) = \sum_{k=0}^{+\infty} p(t - k\Delta T)u(k\Delta T) ,$$

where $p(t)$ denotes the inverse Laplace transform of $P(s)$. Thus, after sampling the output $y(t)$ above, we write

$$y(n\Delta T) = \sum_{k=0}^{+\infty} p\big((n-k)\Delta T\big)u(k\Delta T) ,$$

that is, the discrete-time convolution of the sequences $u(k\Delta T)$ and $p(k\Delta T)$. Thus, $p(k\Delta T)$ represents the impulse response of the exact discretized system.

Remark 6.53 The sampling-holding by a ZOH corresponds to an extrapolation by means of a polynomial of degree zero. If we sample and hold with a more complex converter with transfer function $H_i(s)$, corresponding to an extrapolation by means of a polynomial of higher degree, the previous theorem remains true when $H_0(s)$ is replaced by $H_i(s)$ in the expression (6.57) of $P(s)$. ◇

Theorem 6.52 makes it possible to compute directly the transfer function of the exact discretized of a continuous-time l.c.s. system from the original transfer function, without passing by a state representation. Of course, both methods yield the same results as illustrated on the following example.

Example 6.54 Consider the continuous-time l.c.s. system with transfer function

$$H(s) = \frac{1}{(s+1)(s+2)} . \tag{6.59}$$

We look for the transfer function of its exact discretized at period ΔT. The first method consists in applying Theorem 6.52. For this, we have to evaluate the inverse Laplace transform of the product $P(s) = H(s)H_0(s)$. After decomposition in simple elements, we obtain

$$\mathcal{L}^{-1}\left[P(s)\right](t) = \frac{1}{2}\big(\mathfrak{E}(t) - \mathfrak{E}(t-\Delta T)\big) - \left(e^{-t} - e^{-(t-\Delta T)}\right) + \frac{1}{2}\left(e^{-2t} - e^{-2(t-\Delta T)}\right),$$

where $\mathfrak{E}(t)$ is the Heaviside step function, introduced in Definition 3.11. Therefore, after transformation in z (see Table B.2), we obtain

$$H_d(z) = \frac{(e^{-3\Delta T} + e^{-\Delta T} - 2e^{-2\Delta T}) + z(e^{-2\Delta T} - 2e^{-\Delta T} + 1)}{2(z^2 - z(e^{-\Delta T} + e^{-2\Delta T}) + e^{-3\Delta T})} . \tag{6.60}$$

The second method consists in displaying a state representation of the l.c.s. system with transfer function $H(s)$. We check that the matrices

$$A = \begin{pmatrix} 0 & 1 \\ -2 & -3 \end{pmatrix} , \quad B = \begin{pmatrix} 0 \\ 1 \end{pmatrix} , \quad C = \begin{pmatrix} 1 & 0 \end{pmatrix}$$

constitute a realization of the transfer function $H(s)$, that is, satisfy the relation $H(s) = C(sI - A)^{-1}B$. Now, let us compute the matrices of the state representation of the corresponding discretized system, by applying the results of Theorem 6.2 and the formula (6.49):

$$\begin{cases} A_d = e^{A\Delta T} = \begin{pmatrix} 2e^{-\Delta T} - e^{-2\Delta T} & e^{-\Delta T} - e^{-2\Delta T} \\ -2e^{-\Delta T} + 2e^{-2\Delta T} & -e^{-\Delta T} + 2e^{-2\Delta T} \end{pmatrix} \\ B_d = \int_0^{\Delta T} e^{At} B \, \mathrm{d}t = \begin{pmatrix} (e^{-2\Delta T} - 2e^{-\Delta T} + 1)/2 \\ e^{-\Delta T} - e^{-2\Delta T} \end{pmatrix} . \end{cases}$$

It is straightforward to check that the formula (6.47) giving the transfer matrix of the discrete-time system in state representation provides the expected expression (6.60). △

Remark 6.55 By (6.60), we observe that the poles of the discretized system are $e^{-\Delta T}$ and $e^{-2\Delta T}$ that is, $e^{\lambda_1 \Delta T}$ and $e^{\lambda_2 \Delta T}$, where $\lambda_1 = -1$ and $\lambda_2 = -2$ are the poles of the corresponding continuous-time linear dynamical system. Such a property can be generalized as mentioned in Remark 6.4. ◇

Regarding the zeros of the system, it is clear by (6.60) in the previous example that the discretized system posesses a zero at

$$z = -\frac{e^{-3\Delta T} + e^{-\Delta T} - 2e^{-2\Delta T}}{e^{-2\Delta T} - 2e^{-\Delta T} + 1} ,$$

whereas the continuous-time linear dynamical system with transfer function given by (6.59) has no zero. It is a general observation that the process of discretization adds zeros (see for example [6, Chap. 3]). Thus, a system that is minimum phase in continuous-time can display unstable zeros after discretization.

6.10 Local Stabilization of a Nonlinear Dynamical System by a Control Law in Discrete-Time

Till now, we have synthesized in § 6.7 a control law allowing, under proper conditions, to stabilize a discrete-time controlled and observed linear dynamical system around the origin. We show that, under proper assumptions, the control law discussed in § 6.7, designed for a *discrete-time linear* controlled and observed dynamical system can indeed stabilize an original *continuous-time* controlled and observed *nonlinear* dynamical system.

Consider the model (5.51) discussed in § 5.8, namely the controlled-observed nonlinear dynamical system

$$\begin{cases} \dot{x} = f(x) + g(x)u = f(x) + \sum_{i=1}^{m} g_i(x)u_i \\ \\ y = h(x) = (h_1(x), \dots, h_p(x))^\top . \end{cases} \tag{6.61}$$

Definition 6.56 *Let* (x_E, u_E) *be an equilibrium point of the linear dynamical system* (6.61), *as in Definition 4.34. We call linear controlled-observed* tangent discretized *dynamical system at period* ΔT *of the nonlinear dynamical system* (6.61) *in the neighborhood of the equilibrium point* (x_E, u_E), *the controlled-observed discrete time linear dynamical system*

$$\begin{cases} \xi_{k+1} = A_d \xi_k + B_d v_k \\ \zeta_k = C \xi_k , \end{cases} \tag{6.62}$$

where A_d, B_d *and* C *are the* $n \times n$, $n \times m$ *and* $p \times n$ *matrices given by the following expressions:*

$$\begin{cases} A = \frac{\partial f}{\partial x}(x_E) + \frac{\partial g}{\partial x}(x_E)u_E , \quad B = g(x_E) , \quad C = \frac{\partial h}{\partial x}(x_E) , \\ A_d = e^{\Delta T A} , \quad B_d = \left(\int_0^{\Delta T} e^{tA} \, dt \right) B . \end{cases} \tag{6.63}$$

Remark 6.57 In the more general case where $\dot{x} = f(x, u)$, we refer the reader to the formulas (4.32). ◇

Proposition 6.58 *Suppose that the controlled-observed tangent discretized linear dynamical system* (6.62) *at period* ΔT *of the continuous-time nonlinear dynamical system* (6.61) *is controllable and observable.*

Let K *and* L *be gain matrices such that the matrices* $(A_d - B_d K)$ *and* $(A_d - LC)$ *are asymptotically stable (see Theorem 6.13 and Definition 6.14).*
Then, the closed-loop system

$$\begin{cases} \dot{x} = f(x) + g(x)u \\ y = h(x) \\ \widehat{\xi}_{k+1} = (A_d - B_d K)\widehat{\xi}_k - L\left(C\widehat{\xi}_k - (y(k\Delta T) - h(x_{\mathrm{E}}))\right) \\ u(t) = u_{\mathrm{E}} - K\widehat{\xi}_k \ , \ \ \forall t \in [k\Delta T, (k+1)\Delta T[\end{cases} \tag{6.64}$$

is such that $x(t) \rightarrow_{t \to +\infty} x_{\mathrm{E}}$ *for all initial conditions* $(x(0), \widehat{\xi}_0)$ *in a neighborhood of* $(x_{\mathrm{E}}, 0)$.

Proof Let us denote $\Delta x = x - x_{\mathrm{E}}$ and $\Delta u = u - u_{\mathrm{E}}$. By definition of the matrices A and B in Definition 6.56, we have that

$$f(x_{\mathrm{E}} + \Delta x) + g(x_{\mathrm{E}} + \Delta x)(u_{\mathrm{E}} + \Delta u) = A\Delta x + B\Delta u + \varepsilon_1(\Delta x, \Delta u) \ , \tag{6.65}$$

where ε_1 is negligible with respect to its arguments Δx and Δu in a neighborhood of 0. In the same way, we have that

$$h(x_{\mathrm{E}} + \Delta x) = h(x_{\mathrm{E}}) + C\Delta x + \varepsilon_2(\Delta x) \ , \tag{6.66}$$

where ε_2 is negligible with respect to its arguments Δx in a neighborhood of 0. With these notations, the closed-loop system (6.64) yields, *on the time interval* $[k\Delta T, k\Delta T + \Delta T]$,

$$\begin{cases} \dfrac{d\Delta x(t)}{dt} = A\Delta x(t) - BK\widehat{\xi}_k + \varepsilon_1(\Delta x(t), -K\widehat{\xi}_k) \\ \widehat{\xi}_{k+1} = (A_d - LC - B_d K)\widehat{\xi}_k + L\big(C\Delta x(k\Delta T) + \varepsilon_2(\Delta x(k\Delta T))\big) \ , \end{cases} \tag{6.67}$$

where we used that $y(k\Delta T) = h\big(x_{\mathrm{E}} + \Delta x(k\Delta T)\big)$ by (6.61).

This hybrid dynamical system (in continuous-time and in discrete-time) is stationary (no explicit dependence in k or t), and we study it on the time interval $[0, \Delta T]$. Denote by $\Phi(t, \Delta x_0, \widehat{\xi}_0)$ the solution of the following nonlinear differential equation, for $t \in [0, \Delta T]$:

$$\frac{d\Delta x(t)}{dt} = A\Delta x(t) - BK\widehat{\xi_0} + \varepsilon_1(\Delta x(t), -K\widehat{\xi_0}) \ , \quad \Delta x(0) = \Delta x_0 \ . \tag{6.68}$$

The mapping Φ is continuous, following general results about differential equations [5, 18].

Consider two bounded neighborhoods W_1 and $\widehat{W}_1$ of 0 in $\mathbb{R}^n$. First, we prove the existence of a constant C_1 such that the following inequality holds true:

$$\sup_{t\in[0,\Delta T]} \|\Phi(t, \Delta x_0, \widehat{\xi_0})\| \leq C_1(\|\Delta x_0\| + \|\widehat{\xi_0}\|) \ , \ \forall \Delta x_0 \in W_1 \ , \ \forall \widehat{\xi_0} \in \widehat{W}_1 \ . \tag{6.69}$$

For this purpose, we define

$$W_2 = \{\Phi(t, \Delta x_0, \widehat{\xi_0}) \mid t \in [0, \Delta T] \ , \ \Delta x_0 \in W_1 \ , \ \widehat{\xi_0} \in W_1\} \text{ and } \widehat{W}_2 = W_2 \ .$$

These sets are bounded (because the transformation Φ is continuous) and contain the point 0. As the function ε_1 is negligible with respect to its arguments in a neighborhood of 0, we have that

$$\|\varepsilon_1(\Delta x_0, -K\widehat{\xi_0})\| \leq C_2(\|\Delta x_0\| + \|\widehat{\xi_0}\|) \ , \ \forall \Delta x_0 \in W_2 \ , \ \forall \widehat{\xi_0} \in \widehat{W}_2 \ .$$

Now, taking $\Delta x_0 \in W_1$ and $\widehat{\xi_0} \in \widehat{W}_1$, we have that the solution $\Delta x(t) = \Phi(t, \Delta x_0, \widehat{\xi_0})$ of (6.68) belongs to $\widehat{W}_2 = W_2$, by definition of these sets and, therefore, the last term in (6.68) satisfies

$$\|\varepsilon_1(\Delta x(t), -K\widehat{\xi_0})\| \leq C_2(\|\Delta x(t)\| + \|\widehat{\xi_0}\|) \ . \tag{6.70}$$

If $\Delta x_0 \in W_1$ and $\widehat{\xi_0} \in \widehat{W}_1$, we deduce from (6.68) and (6.70) the following inequality

$$\|\Delta x(t)\| \leq (\|A\| + C_2) \int_0^t \|\Delta x(s)\| + \Delta T(\|BK\| + C_2)\|\widehat{\xi_0}\| + \|\Delta x_0\| \ , \ \forall t \in [0, \Delta T] \ ,$$

and (6.69) is a consequence of the Gronwall Lemma (see [18, p.117]).

Now, let us focus on the transitions between the times $k\Delta T$ and $k\Delta T + \Delta T$ for the closed-loop system (6.64). From (6.67), we deduce that

$$\begin{aligned}
\Delta x(k\Delta T + \Delta T) &= e^{\Delta T A}\Delta x(k\Delta T) \\
&\quad + \int_{k\Delta T}^{k\Delta T+\Delta T} e^{(k\Delta T+\Delta T-s)A}\left(-BK\widehat{\xi_k} + \varepsilon_1(\Delta x(s), -K\widehat{\xi_k})\right) ds \\
&= e^{\Delta T A}\Delta x(k\Delta T) \\
&\quad + \int_0^{\Delta T} e^{sA}\left(-BK\widehat{\xi_k} + \varepsilon_1(\Delta x(k\Delta T + \Delta T - s), -K\widehat{\xi_k})\right) ds \\
&= A_d\Delta x(k\Delta T) - B_d K\widehat{\xi_k}
\end{aligned}$$

$$
\begin{aligned}
&+\int_0^{\Delta T} e^{sA}\varepsilon_1(\Delta x(k\Delta T+\Delta T-s),-K\widehat{\xi_k})\,\mathrm{d}s \\
&=(A_d-B_dK)\Delta x(k\Delta T)-B_dK(\widehat{\xi_k}-\Delta x(k\Delta T)) \\
&\quad+\int_0^{\Delta T} e^{sA}\varepsilon_1(\Delta x(k\Delta T+\Delta T-s),-K\widehat{\xi_k})\,\mathrm{d}s\,.
\end{aligned}
$$

By (6.67) and the previous equation, the vector

$$
\Delta X_k=\begin{pmatrix}\Delta x(k\Delta T)\\ \widehat{\xi_k}-\Delta x(k\Delta T)\end{pmatrix} \tag{6.71}
$$

satisfies the induction

$$
\begin{aligned}
\Delta X_{k+1}&=\begin{pmatrix}A_d-B_dK & -B_dK\\ 0 & A_d-LC\end{pmatrix}\Delta X_k \\
&+\begin{pmatrix}\displaystyle\int_0^{\Delta T} e^{sA}\varepsilon_1(\Delta x(k\Delta T+\Delta T-s),-K(\Delta e_k+\Delta x(k\Delta T)))\,\mathrm{d}s\\ \displaystyle L\varepsilon_2(\Delta x(k\Delta T))-\int_0^{\Delta T} e^{sA}\varepsilon_1(\Delta x(k\Delta T+\Delta T-s),-K(\Delta e_k+\Delta x(k\Delta T)))\,\mathrm{d}s\end{pmatrix}.
\end{aligned}
$$

Then, let us introduce the notations

$$
F=\begin{pmatrix}A_d-B_dK & -B_dK\\ 0 & A_d-LC\end{pmatrix},
$$

for the linear part, and

$$
\mathfrak{p}\begin{pmatrix}\eta\\ \rho\end{pmatrix}=\begin{pmatrix}\displaystyle\int_0^{\Delta T} e^{sA}\varepsilon_1\big(\Phi(\Delta T-s,\rho,\eta+\rho),-K(\eta+\rho)\big)\,\mathrm{d}s\\ \displaystyle L\varepsilon_2(\rho)-\int_0^{\Delta T} e^{sA}\varepsilon_1\big(\Phi(\Delta T-s,\rho,\eta+\rho),-K(\eta+\rho)\big)\,\mathrm{d}s\end{pmatrix},
$$

for the perturbation term. By the stationarity of the system (6.67), we check that

$$
\Delta X_{k+1}=F\Delta X_k+\mathfrak{p}(\Delta X_k)\,,\ \ \forall k\in\mathbb{N}\,. \tag{6.72}
$$

Now, by the estimation (6.69), the perturbation term $\mathfrak{p}$ is, as ε_1 in (6.65) and ε_2 in (6.66), negligible with respect to its arguments in a neighborhood of 0. The matrix F being asymptotically stable by assumption, we deduce from Proposition 6.17 that $\|\Delta X_k\|\to_{k\to+\infty}0$ for initial conditions $(x(0),\widehat{\xi_0})$ in a neighborhood of $(x_{\mathrm{E}},0)$.

Now, for $t\in[k\Delta T,(k+1)\Delta T[$, we have that,

$$
\|\Delta x(t)\|\le\sup_{s\in[k\Delta T,k\Delta T+\Delta T]}\Phi(s,\Delta x(k\Delta T),\widehat{\xi_k})\le C_1(\|\Delta x_k\|+\|\widehat{\xi_k}\|)\,,
$$

because the hybrid dynamical system (6.67) is stationary, and by the estimation (6.69). By the definition (6.71) of ΔX_k, and by $\|\Delta X_k\| \to_{k\to+\infty} 0$, we deduce that

$$\|x(t) - x_{\mathrm{E}}\| = \|\Delta x(t)\| \to_{t\to+\infty} 0 ,$$

for initial conditions $(x(0), \widehat{\xi_0})$ in a neighborhood of $(x_{\mathrm{E}}, 0)$. ◇

Let us underline that, unlike the case of a controlled-observed *linear* system, such a control law is a priori stabilizing only when the state x is in a *neighborhood* of the equilibrium point x_{E}.

6.11 Practical Set Up

In this chapter, we have brought an answer to the problem raised by the discrete-time character of controls and of measurements, and discussed in § 5.10. The question of the modes placement remains open because of the stability-precision dilemma yet discussed in § 5.10: the sensitivity analysis with respect to parameters performed for continuous-time systems remains valid and can be extended without difficulty to the observer-regulator synthesis case obtained by the separation principle. Also, the consideration of perturbations has not been tackled in the multivariable case.

In the following chapter, we bring a possible answer to the question of the modes placement. The same framework also contributes to tackle the issue of how to deal with perturbations.

6.12 Exercises

Exercise 6.12.1 We proceed with the study of the mixing process begun in Exercises 2.6.1 and 5.11.1.

1. Compute the exact discretized (with a zero-order hold) of the tangent linear dynamical system, sampled at period ΔT.
2. Study the stability of the discrete-time system obtained, as well as its controllability and observability properties.

Exercise 6.12.2 Same questions with the system ball + rail discussed in Exercises 2.6.3, 47.8.5 and 5.11.3.

Exercise 6.12.3 Consider the continuous-time linear dynamical system:

$$\begin{cases} \begin{pmatrix} \dot{x}_1 \\ \dot{x}_2 \end{pmatrix} = \begin{pmatrix} 0 & 1 \\ -1 & 0 \end{pmatrix} \begin{pmatrix} x_1 \\ x_2 \end{pmatrix} + \begin{pmatrix} 0 \\ 1 \end{pmatrix} u \\ y = \begin{pmatrix} 1 & 0 \end{pmatrix} \begin{pmatrix} x_1 \\ x_2 \end{pmatrix} . \end{cases}$$

1. Show that this system is controllable and observable.
2. Compute its exact discretized with a ZOH at sampling period ΔT.
3. What is the condition on ΔT for the exact discretized system to remain observable and controllable?

Exercise 6.12.4 Calculate the z-transform of the following signals sampled at the period ΔT: e^{at}, $\alpha \cos \omega t$, $\alpha \sin \omega t$ (see § B.2).

Exercise 6.12.5 Prove the final value Theorem Z6 in § B.2, that is,

$$\lim_{z \to 1} (1 - z^{-1}) X(z) = \lim_{k \to +\infty} x_k \ ,$$

if $X(z)$ is the z-transform of the sequence (x_k).

Exercise 6.12.6 Show that the transfer function

$$H(z) = 1 - 2z^{-1} \cos \omega \Delta T + z^{-2}$$

asymptotically filters the periodic inputs of pulsation ω.

Chapter 7
Quadratic Optimization and Linear Filtering

7.1 Introduction

Up to now, we have shed light on control methods which require the modes placement of the matrix of the closed-loop system in a stability domain, whether an open half-space in continuous-time or an open disk in discrete-time. However, no recommendation is given regarding the positioning of these modes: should they be close to the border? or far away?

We can extend the analysis in § 5.10 for continuous-time dynamical systems, and deduce that, for a linear dynamical system in discrete-time, a compromise must be achieved. Indeed, modes close to the unit circle provide stability but poor precision, whereas when close to the circle center, precision is high but stability is poor. This compromise can be achieved by tuning the parameters of the control law, namely the gains of the control and of the observer, and performing numerical simulations. We can also try to quantify stability and precision and weigh them in a criterion (objective function) to be minimized with respect to the "best control". This is the program that we implement in § 7.2 where we show how the gains of the controller can be obtained by means of the solution of a quadratic optimization problem.

In a dual way, the issue for the observer is one of quantifying stability—here the capacity to absorb measurement errors—and precision, the discrepancy between the estimation of the state and the state itself. This question is tackled in § 7.3 where we show how the gain of the observer can be obtained by means of the solution of a linear filtering problem, after introducing a probabilistic framework to account for perturbations and measurement errors.

We detail the results in the discrete-time case, and we state them more briefly in the continuous-time case in § 7.4.

B. d'Andréa-Novel and M. De Lara, *Control Theory for Engineers*,
DOI: 10.1007/978-3-642-34324-7_7, © Springer-Verlag Berlin Heidelberg 2013

7.2 Quadratic Optimization and Controller Modes Placement

Consider the controlled linear dynamical system in discrete-time

$$x_{k+1} = A_d x_k + B_d u_k \ , \quad k \in \mathbb{N} \, . \tag{7.1}$$

Our goal is to drive the state towards 0. By the proof of Theorem 6.21, we know that this can be achieved by a sequence of n controls as soon as the couple (A_d, B_d) is controllable. However, we favor here another approach that takes into account the link between the "cost" of a control (the energy to exert it) and its impact on the state (the distance to the target state 0), independent of any controllability assumption. We touch on how to quantify this cost and this impact, and we define a benchmark to compare the performances of different controls. We show the existence of a control law which is *optimal* regarding this criterion. This control law is a linear state feedback, *but with a gain that depends on time*. This result is obtained without assumption of controllability. When controllability holds true, we show that, as time goes on, this gain converges towards a fixed gain that makes the matrix of the closed-loop dynamics asymptotically stable. Therefore, we now have a method of mode placement of the regulator, and the linear feedback (6.31) of Theorem 6.23 is thus recovered as solution of an optimization problem.

We suppose given R, symmetric *definite* positive matrix of size m (number of scalar controls), and Q symmetric positive matrix of size n (dimension of the state).

We refer the reader to Appendix A for recalls on symmetric matrices, and for results on Riccati equations.

7.2.1 Optimization in Finite Horizon

Here, the finite horizon is an integer $f \geq 1$. For a sequence of controls $u_0, \ldots, u_{f-1}$ and an initial state x_0, we obtain a trajectory $x_0, x_1, \ldots, x_{f-1}, x_f$ by (7.1).

We quantify cost and impact of this sequence of controls in a quadratic way within an *intertemporal criterion* as follows.

Definition 7.1 *If Q_f is a symmetric positive matrix of size n, we define the* intertemporal criterion *(with* penalization *on the final state) by*

$$J(u) = J(u_0, \ldots, u_{f-1}) = \sum_{k=0}^{f-1} (u_k^\top R u_k + x_k^\top Q x_k) + x_f^\top Q_f x_f \ , \tag{7.2}$$

where the state trajectory $x_0, x_1, \ldots, x_{f-1}, x_f$ is given by (7.1). *The symmetric positive matrices Q and R are called* ponderation matrices.

Notice that the criterion J is nonnegative, and that it weighs both the state and the control. According to the relative importance of the ponderation matrices Q and R, more or less weight is attached to the state or to the control in the criterion J.

Then, a sequence of controls u_0, ..., u_{f-1} that minimizes the criterion J is looked after. The following proposition provides a solution.

Proposition 7.2 *The following backward induction is well defined:*

$$\begin{cases} S_f &= Q_f \\ S_{k-1} &= Q + A_d^\top S_k A_d - A_d^\top S_k B_d (R + B_d^\top S_k B_d)^{-1} B_d^\top S_k A_d \\ k &= f, \ldots, 1 \, . \end{cases} \tag{7.3}$$

The solution sequence S_f,..., S_0 consists of symmetric positive matrices. The minimum of the criterion (7.2) *is*

$$\min_u J(u) = x_0^\top S_0 x_0 \, . \tag{7.4}$$

Denoting

$$K_k = (R + B_d^\top S_{k+1} B_d)^{-1} B_d^\top S_{k+1} A_d \, , \ k = 0, \ldots, f-1, \tag{7.5}$$

*this minimum is achieved for the following linear state feedback (*optimal control*)*

$$u_k^\star = -K_k x_k^\star \, , \ k = 0, \ldots, f-1, \tag{7.6}$$

where $(x_k^\star)_{k=0,\ldots,f-1}$ is the optimal trajectory *corresponding to the application of the optimal control* (7.6) *to the dynamics* (7.1), *namely $x_0^\star = x_0$ and*

$$x_{k+1}^\star = A_d x_k^\star + B_d u_k^\star = (A_d - B_d K_k) x_k^\star \, , \ k = 0, \ldots, f-1.$$

Proof For the backward induction (7.3) to be well defined, it suffices to show that every matrix $R + B_d^\top S_k B_d$ ($k = f$, ..., 1) is invertible.

Since $S_f = Q_f \geq 0$, then, by assumption on R (positive *definite*), we have that

$$R + B_d^\top S_f B_d \geq R > 0 \, ,$$

which implies that the matrix $R + B_d^\top S_f B_d$ is invertible. Therefore, the expression (7.3) has proper meaning for $k = f$, and S_{f-1} is well defined. The backward induction (7.3) is well defined as soon as we show that

$$S \geq 0 \Rightarrow Q + A_d^\top S A_d - A_d^\top S B_d (R + B_d^\top S B_d)^{-1} B_d^\top S A_d \geq 0 \, .$$

Now, with the notations of Definition A.1 in Appendix A, we write

$$\psi(S) = Q + A_d^\top S A_d - A_d^\top S B_d (R + B_d^\top S B_d)^{-1} B_d^\top S A_d \, ,$$

as in (A.1), with $Q = H^\top H$ (H is a square root of Q), $F = A_d$ and $G = B_d$. By Lemma A.2, we obtain that $\psi(S) \geq \psi(0) = Q \geq 0$ since $S \geq 0$.

Now, we are going to show that the criterion $J(u)$ in (7.2) can be written under the form

$$J(u) = x_0^\top S_0 x_0 + \sum_{k=1}^{f} (u_{k-1} + K_{k-1}x_{k-1})^\top (R + B_d^\top S_k B_d)(u_{k-1} + K_{k-1}x_{k-1}) \,, \quad (7.7)$$

from which the rest of the proposition is immediately deduced.

We write the criterion (7.2) under the form:

$$\begin{aligned} J(u) &= \sum_{k=0}^{f-1} (u_k^\top R u_k + x_k^\top Q x_k) + x_f^\top Q_f x_f \\ &= \sum_{k=1}^{f} \begin{pmatrix} u_{k-1}^\top & x_{k-1}^\top \end{pmatrix} \begin{pmatrix} R & 0 \\ 0 & Q \end{pmatrix} \begin{pmatrix} u_{k-1} \\ x_{k-1} \end{pmatrix} + x_f^\top Q_f x_f \,. \end{aligned}$$

By (7.3), we have that

$$x_f^\top Q_f x_f = x_f^\top S_f x_f = x_0^\top S_0 x_0 + \sum_{k=1}^{f} (x_k^\top S_k x_k - x_{k-1}^\top S_{k-1} x_{k-1}) \,,$$

where, by (7.1), $x_k^\top S_k x_k$ is given by the expression

$$\begin{aligned} x_k^\top S_k x_k &= (A_d x_{k-1} + B_d u_{k-1})^\top S_k (A_d x_{k-1} + B_d u_{k-1}) \\ &= \begin{pmatrix} u_{k-1}^\top & x_{k-1}^\top \end{pmatrix} \begin{pmatrix} B_d^\top S_k B_d & B_d^\top S_k A_d \\ A_d^\top S_k B_d & A_d^\top S_k A_d \end{pmatrix} \begin{pmatrix} u_{k-1} \\ x_{k-1} \end{pmatrix} . \end{aligned}$$

Combining the three above expressions, we obtain:

$$\begin{aligned} J(u) = x_0^\top S_0 x_0 + \sum_{k=1}^{f} &\begin{pmatrix} u_{k-1}^\top & x_{k-1}^\top \end{pmatrix} \\ &\times \begin{pmatrix} R + B_d^\top S_k B_d & B_d^\top S_k A_d \\ A_d^\top S_k B_d & Q + A_d^\top S_k A_d - S_{k-1} \end{pmatrix} \begin{pmatrix} u_{k-1} \\ x_{k-1} \end{pmatrix} . \end{aligned}$$

Using the induction (7.3), we observe that the lower right term of the last matrix is

$$Q + A_d^\top S_k A_d - S_{k-1} = A_d^\top S_k B_d (R + B_d^\top S_k B_d)^{-1} B_d^\top S_k A_d \,.$$

Computation shows that

$$J(u) = x_0^\top S_0 x_0 + \sum_{k=1}^{f} (u_{k-1} + (R + B_d^\top S_k B_d)^{-1} B_d^\top S_k A_d x_{k-1})^\top (R + B_d^\top S_k B_d)$$
$$(u_{k-1} + (R + B_d^\top S_k B_d)^{-1} B_d^\top S_k A_d x_{k-1}) ,$$

and (7.7) is proven thanks to the expression (7.5). □

REMARKS

- If the matrices A_d, B_d, Q and R depend on the time k, the result and the previous formulas remain valid. This makes it possible to compute optimal linear state feedbacks, not stationary and not necessarily stabilizing, around a trajectory (see § 5.9.3).
- Notice, however, that even with stationary matrices A_d, B_d, Q and R, the control that minimizes the criterion J is a linear state feedback (7.6) with *time varying gain matrix* K_k given by (7.5).

Now, we show that the solution of an optimization problem in infinite horizon is a linear state feedback control displaying *fixed gain*.

7.2.2 Optimization in Infinite Horizon. Links with Controllability

We now show that the stabilizing linear state feedback control (6.31) with stationary gains discussed in Chap. 6 minimizes an intertemporal criterion.

Definition 7.3 *Let $\mathcal{U}$ be the set of control sequences $u = (u_k)_{k\in\mathbb{N}}$ such that the state trajectory $(x_k)_{k\in\mathbb{N}}$ given by (7.1) converges towards 0 and that $\sum_{k=0}^{+\infty}(u_k^\top R u_k + x_k^\top Q x_k) < +\infty$. We define an intertemporal criterion J on $\mathcal{U}$ as follows:*

$$J(u) = \sum_{k=0}^{+\infty} (u_k^\top R u_k + x_k^\top Q x_k) , \quad \forall u \in \mathcal{U} . \tag{7.8}$$

Here is a first sufficient condition of existence of an optimal control sequence for the criterion J.

Proposition 7.4 *Suppose that there exists a symmetric positive matrix S such that:*

1. *S is solution of the* algebraic stationary Riccati equation

$$S = Q + A_d^\top S A_d - A_d^\top S B_d (R + B_d^\top S B_d)^{-1} B_d^\top S A_d ; \tag{7.9}$$

2. *the matrix $A_d - B_d(R + B_d^\top S B_d)^{-1} B_d^\top S A_d$ is asymptotically stable.*

Then, the minimum of the criterion (7.8) *on* $\mathcal{U}$ *satisfies*

$$\min_{u \in \mathcal{U}} J(u) = x_0^\top S x_0 \; . \tag{7.10}$$

Denoting

$$K = (R + B_d^\top S B_d)^{-1} B_d^\top S A_d \; , \tag{7.11}$$

this minimum is achieved for the following control sequence, given by linear state feedback,

$$u_k^\star = -K x_k^\star \; , \; k \in \mathbb{N} \; , \tag{7.12}$$

where $(x_k^\star)_{k \in \mathbb{N}}$ *is the optimal trajectory corresponding to the application of the optimal control* (7.12)*, that is,* $x_0^\star = x_0$ *and*

$$x_{k+1}^\star = A_d x_k^\star + B_d u_k^\star = (A_d - B_d K) x_k^\star \; , \; k \in \mathbb{N} \; . \tag{7.13}$$

Proof For all integer $f \geq 1$, denote by

$$J_f(u) = \sum_{k=0}^{f-1} (u_k^\top R u_k + x_k^\top Q x_k)$$

the criterion up to time f, without penalization of the final state, that is, $Q_f = 0$ in (7.2). Using (7.9), a computation similar to that of the proof of Proposition 7.2 makes it possible to write

$$J_f(u) + x_f^\top S x_f = x_0^\top S x_0 + \sum_{k=1}^{f} (u_{k-1} + K x_{k-1})^\top (R + B_d^\top S B_d)(u_{k-1} + K x_{k-1}) \; .$$

We deduce that

$$J_f(u) + x_f^\top S x_f \geq x_0^\top S x_0 = x_0^{\star\top} S x_0^\star = J_f(u^\star) + x_f^{\star\top} S x_f^\star \; , \tag{7.14}$$

where the control $u^\star$ is given by (7.12). We have that $u^\star$ belongs to $\mathcal{U}$ because, by asssumption, the matrix $A_d - B_d K$ is asymptotically stable, so that (7.12) and (7.13) yield state and control trajectories that converge exponentially to zero (see the proof of Proposition 6.12).

For $u \in \mathcal{U}$, we have that:

$$\begin{aligned} J(u) &= \lim_{f \to +\infty} J_f(u) \text{ by definition of } \mathcal{U} \\ &= \lim_{f \to +\infty} (J_f(u) + x_f^\top S x_f) \text{ because } x_f \to_{f \to +\infty} 0 \text{ by definition of } \mathcal{U} \end{aligned}$$

$$
\begin{aligned}
&\geq \lim_{f \to +\infty} (J_f(u^\star) + x_f^{\star\top} S x_f^\star) \text{ by (7.14)} \\
&= \lim_{f \to +\infty} J_f(u^\star) \text{ because } x_f^\star \to_{f \to +\infty} 0 \text{ since } u^\star \in \mathcal{U} \\
&= J(u^\star) .
\end{aligned}
$$

This ends the proof. □

The existence of a matrix S that satisfies the two assumptions of Proposition 7.4 is assured under the following assumptions on the dynamics matrix A_d, the control matrix B_d and the state ponderation matrix Q (the proof is given in Proposition A.4).

Proposition 7.5 *Suppose that*

1. *the couple* (A_d, B_d) *is controllable;*
2. *there exists a square root* H *of the matrix* Q $(Q = H^\top H)$ *such that the couple* (A_d, H) *(or* (A_d, HA_d)*) is observable.*

Then, there exists a unique symmetric positive matrix S *which satisfies the two assumptions of Proposition 7.4. Moreover, the matrix* S *is* positive definite *and is the limit of all the sequences*

$$
S_{k+1} = Q + A_d^\top S_k A_d - A_d^\top S_k B_d (R + B_d^\top S_k B_d)^{-1} B_d^\top S_k A_d , \tag{7.15}
$$

for all initial condition $S_0 \geq 0$.

7.2.3 Implementation

Propositions 7.4 and 7.5 offer an alternative to the method of modes placement of Theorem 6.23. Indeed, if the system (7.1) is controllable and if we choose ponderation matrices of the form $Q = H^\top H$ and R such that the couple (A_d, H) (or (A_d, HA_d)) is observable, then the unique symmetric positive solution S of the algebraic Riccati equation (7.9) makes it possible to compute the gain K in (7.11), which makes the matrix $A_d - B_d K$ asymptotically stable.

From a practical point of view, we approximate S by a matrix S_k given by the induction (7.15) and, if k is large enough, the gain matrix K_k in (7.5) is close to K. Therefore, $A_d - B_d K_k$ is asymptotically stable for this fixed k by Proposition 2.28.

We can note that the problem of the modes choice is now an issue of choosing ponderation matrices Q and R in a quadratic intertemporal criterion.

7.3 Kalman-Bucy Filter and Observer Modes Placement

For a linear controllable system, an appropriate linear *state* feedback makes the closed-loop dynamics matrix asymptotically stable (see Theorem 6.23). Now, the state x_k is often only partially known through measurements and we discussed in

Theorem 6.27 how to build a dynamical system driven by the measurements y_1, ..., y_k, the output of which is an approximation $\hat{x}_k$ of the state x_k. In a sense, this observer performs a compromise between the information about the state brought by the measurements (more or less corrupted by measurements errors) and the prevision of the state produced by the dynamics (6.3) (more or less faithful to the theoretical future state). In what follows, we propose to quantify the quality of measurements and of the state-model by means of *probabilistic* notions, and to find the observer as a *conditional expectation*, that is, one that minimizes a quadratic criterion, weighting the random effects.

We suppose that the reader is familiar with basic notions of probability calculus: random variable, expectation, variance, dispersion, conditional expectation, independence [11,31]. Some review of Gaussian vectors is given in Appendix C.

The introduction of a *probabilistic*, or *stochastic* framework is a way to handle measurement errors, rather inevitable due to the nature of sensors. At each time k, we do not measure $y_k = h\big(x(k\Delta T)\big)$ as in (6.60), but rather

$$y_k = h\big(x(k\Delta T)\big) + \text{measurement noise.}$$

The quality of the result supplied by the observers in Theorems 6.27 and 6.28 depends on the importance of the "measurement noise," assumed to be additive here, according to the sensors quality. How to quantify this importance? By nature, this noise is not known at each measurement instant, but some general information is assumed:

- if the sensor is not biaised, this noise is zero in the mean;
- the amplitude of the noise is an indicator of the precision of the sensor.

We model such a noise as a *random variable*, or as *random vector*. Thus, the value y_k of the measurement becomes itself random since

$$y_k = h\big(x(k\Delta T)\big) + \text{random vector.}$$

Moreover, the introduction of random variables is a way to capture other effects, such as those due to linearization. Indeed, the observer is generally built from a first-order approximation characterized by $C = \frac{\partial h}{\partial x}(x_{\mathrm{E}})$ (see Definition 6.56). We have that

$$y_k = h(x_{\mathrm{E}}) + C(x(k\Delta T) - x_{\mathrm{E}}) + \text{higher order terms} + \text{noise,}$$

and it is tempting to gather the two last terms in a random vector. This is why, rather than the tangent discretized linear dynamical system (6.61), we consider the representation

$$\begin{cases} \xi_{k+1} = A_d \xi_k + B_d \upsilon_k \\ \zeta_k = C\xi_k + w_k \end{cases}$$

where $w_0, w_1, \ldots$ is a sequence of random vectors (with the same dimension as that of the observations). Statistical properties of this random sequence capture and quantify the effets of linearization and of measurement noise. But, before highlighting this issue, one may wonder why only the observation is corrupted by noise, whereas the dynamics would be immune to it. We consider randomness in the dynamics, because the linear dynamics of the state is also often an approximation due to linearization, and because of perturbations too. This is why, we now focus on systems defined as below.

Definition 7.6 *We call* controlled and observed linear dynamical system *in discrete-time with* additive noises *the following system*

$$\begin{cases} x_{k+1} = A_d x_k + B_d u_k + v_k \ , \ k \geq 0 \\ y_k = C x_k + w_k \ , \ k \geq 1 \ , \end{cases} \tag{7.16}$$

where the sequence of random vectors $(v_k)_{k \in \mathbb{N}}$ *is called* state noise *and* $(w_k)_{k \geq 1}$ measurement noise.

Remark 7.7 We also use the terminology of *linear system excited* by the state noise $(v_k)_{k \in \mathbb{N}}$. ◇

For such systems, the problem of the state estimation by means of measurements is formulated as a problem of *filtering* and rests upon the probabilistic notions of distribution and of conditional expectation: characterize, for all time k, the *conditional distribution* of the state x_k knowing the observations $y_1, \ldots, y_k$ up to time k. This problem can be solved explicitly in the case of Gaussian linear systems (to be discussed below, see also [41]).

7.3.1 The Kalman-Bucy Filter

The observer displayed in Theorem 6.28 is an algorithm allowing to recursively compute $\hat{x}_{k+1}$ in function of $\hat{x}_k$ (state of the dynamical system observer) and of y_{k+1} (input that drives the observer dynamical system). This recursive structure is also that of the solution brought by Kalman and Bucy to the problem of estimation of the state of a linear Gaussian dynamical system.

Definition 7.8 *We say that a sequence of random vectors is a (stationary) Gaussian* white noise *if it is an independent (and identically distributed)* sequence of centered random vectors *having Gaussian distribution.*

Remark 7.9 In this definition, "noise" corresponds to centered or of zero mean, "white" corresponds to decorrelation/independence ("colored noise" corresponds to correlation), and "stationary" to identically distributed. ◇

Definition 7.10 *The system* (7.16) *is said to be* linear Gaussian dynamical system *with* independent noises *if*

1. *the control* u_k *depends* affinely *on the observations* $y_1, \ldots, y_k$ *up to time* k;
2. *the initial state* x_0 *is a Gaussian random vector;*

3. *the sequence* $(v_k)_{k\in\mathbb{N}}$ *of state noises is a stationary Gaussian white noise;*
4. *the sequence* $(w_k)_{k\geq 1}$ *of observation noises is a stationary Gaussian white noise;*
5. *the random vectors and processes* x_0, $(v_k)_{k\in\mathbb{N}}$ *and* $(w_k)_{k\geq 1}$ *are independent.*

Consider the linear Gaussian system (7.16) with independent noises. We set

$$\hat{x}_k := \mathbb{E}\big[x_k \mid y_1, \ldots, y_k\big] , \tag{7.17}$$

the conditional mean $\hat{x}_k$ given by the conditional expectation of the state x_k knowing the observations y_1, ..., y_k up to time k. We also set

$$P_k := \mathsf{D}(\hat{x}_k - x_k) , \tag{7.18}$$

the dispersion matrix of the error

$$e_k := \hat{x}_k - x_k . \tag{7.19}$$

The *innovation* at time $k + 1$ is the centered random vector

$$i_{k+1} := y_{k+1} - \mathbb{E}\big[y_{k+1} \mid y_1, \ldots, y_k\big] . \tag{7.20}$$

Proposition 7.11 *Consider the linear Gaussian system* (7.16) *with independent noises. Then, the conditional distribution of the state* x_k *knowing the observations* y_1,..., y_k *up to time k is Gaussian with mean* $\hat{x}_k$ *given by* (7.17) *and dispersion matrix* P_k *given by* (7.18). *Moreover, the random vector* $(i_{k+1}, y_1, \ldots, y_k)$ *is Gaussian, and the innovation* i_{k+1} *is independent of the outputs* y_1,..., y_k.

Proof First, we show that, for all integer k, the random vector $(x_k, y_k, \ldots, y_1)$, made of "state-observations" vectors, is Gaussian.

Indeed, the random vector $(x_0, v_0, v_1, \ldots, v_k, w_1, \ldots, w_k)$ is Gaussian by Proposition C.9. By Proposition C.7, the random vector $(x_k, y_k, \ldots, y_1)$ is Gaussian since it is obtained by an affine transformation from x_0, v_0, v_1,..., v_k, w_1,..., w_k, as can be seen on the expression (7.16), and by the property 1 in Definition 7.10 that the control u_k depends *affinely* on the observations y_1,..., y_k.

At last, thanks to Proposition C.11, we conclude that the conditional distribution of the state x_k knowing the observations y_1,..., y_k up to time k is Gaussian, with the parameters $\hat{x}_k$ and P_k.

By Proposition C.11, the random vector $(i_{k+1}, y_1, \ldots, y_k)$ is Gaussian and the innovation i_{k+1} is independent of the outputs y_1,..., y_k. □

The following so-called *Kalman-Bucy filter* makes it possible to *recursively* compute the parameters $\hat{x}_k$ and P_k of the conditional distribution of the state x_k by *induction*.

We set the so-called *predicted state*

$$\hat{x}_{k+1}^{-} := \mathbb{E}\big[x_{k+1} \mid y_1, \ldots, y_k\big] , \tag{7.21}$$

and the following matrices

$$\begin{cases} P_{k+1}^{-} = \mathsf{D}(x_{k+1} - \hat{x}_{k+1}^{-}) \\ L_{k+1} = P_{k+1}^{-} C^{\top} (C P_{k+1}^{-} C^{\top} + R)^{-1} . \end{cases} \tag{7.22}$$

Theorem 7.12 *Consider the linear Gaussian system* (7.16) *with independent noises. Suppose that*

- *the initial state is Gaussian with mean* $\mathbb{E}\big[x_0\big] = \overline{x}_0$ *and dispersion* $\mathsf{D}(x_0) = \overline{P}_0$*;*
- *the state stationary Gaussian white noise has dispersion* $\mathsf{D}(v_k) = Q$, $\forall k \in \mathbb{N}$*;*
- *the measurement stationary Gaussian white noise has dispersion* $\mathsf{D}(w_k) = R$, $\forall k \geq 1$, *where* $R > 0$.

Then, the characteristics of the conditional distribution of the state x_k *knowing the observations* y_1,..., y_k, *namely the vector* $\hat{x}_k$ *in* (7.17) *and the dispersion matrix* P_k *in* (7.18)*, are given by the following recursive algorithm, called* Kalman-Bucy filter*:*

initialization: $\hat{x}_0 = \overline{x}_0 \ , \ \ P_0 = \overline{P}_0$

$$\begin{aligned} &\textit{propagation:} && \begin{cases} \hat{x}_{k+1}^{-} = A_d \hat{x}_k + B_d u_k \\ P_{k+1}^{-} = A_d P_k A_d^{\top} + Q \end{cases} \\ &\textit{updating:} && \begin{cases} \hat{x}_{k+1} = \hat{x}_{k+1}^{-} + L_{k+1}(y_{k+1} - C\hat{x}_{k+1}^{-}) \\ L_{k+1} = P_{k+1}^{-} C^{\top} (C P_{k+1}^{-} C^{\top} + R)^{-1} \\ P_{k+1} = (I - L_{k+1} C) P_{k+1}^{-} . \end{cases} \end{aligned} \tag{7.23}$$

Remark 7.13 The propagation step in the previous induction only makes use of the observations y_1, ..., y_k:

$$\hat{x}_{k+1}^{-} = \mathbb{E}\big[x_{k+1} \mid y_1, \ldots, y_k\big] \text{ and } P_{k+1}^{-} = \mathsf{D}(x_{k+1} - \hat{x}_{k+1}^{-}) .$$

The updating step takes into account the contribution of the last observation y_{k+1}. $\diamond$

The proof of Theorem 7.12 is classical [41]. The sketch of the proof is the following:

- because $(x_{k+1}, y_1, \ldots, y_{k+1})$ is a Gaussian vector, the conditional expectation $\mathbb{E}\big[x_{k+1} \mid y_1, \ldots, y_{k+1}\big]$ of x_{k+1} knowing y_1,..., y_{k+1} coincides with the orthogonal projection of x_{k+1} on the linear space $\mathbb{Y}_{k+1}$ of random variables generated by $1, y_1, \dots, y_{k+1}$;

- this space can be written as a direct orthogonal sum $\mathbb{Y}_{k+1} = \mathbb{Y}_k \overset{\perp}{\oplus} \mathbb{R}i_{k+1}$ where i_{k+1} is the innovation defined in (7.20);
- the orthogonal projection of x_{k+1} on $\mathbb{Y}_{k+1}$ is expressed as the sum of the orthogonal projection of x_{k+1} on $\mathbb{Y}_k$ and on $\mathbb{R}i_{k+1}$;
- the conditional expectation $\mathbb{E}\big[x_{k+1} \mid y_1, \dots, y_{k+1}\big]$ is a sum, with one term that only depends on $y_1, \dots, y_k$ and the other one only on the innovation i_{k+1}.

In the new observation y_{k+1} at time $k+1$, the innovation represents the part which is decorrelated (even independent) from the previous observations $y_1, \dots, y_k$ up to time k. This term is crucial to obtain a recursive formula.

Now, we skip to the proof of Theorem 7.12.

Proof (Theorem 7.12)

We compute

$$\begin{aligned}
\hat{x}_{k+1} &= \mathbb{E}\big[x_{k+1} \mid y_1, \dots, y_{k+1}\big] \\
&= \mathbb{E}\big[x_{k+1} \mid y_1, \dots, y_k, i_{k+1}\big] \text{ by (7.20)} \\
&= \mathbb{E}\big[x_{k+1} \mid y_1, \dots, y_k\big] + L_{k+1} i_{k+1} \text{ by Lemma C.12,}
\end{aligned}$$

where $\mathsf{cov}(x_{k+1}, i_{k+1}) = L_{k+1}\mathsf{D}(i_{k+1})$.

By (7.21), we have that

$$\begin{aligned}
\hat{x}_{k+1}^{-} &= \mathbb{E}\big[A_d x_k + B_d u_k + v_k \mid y_1, \dots, y_k\big] \text{by (7.1)} \\
&= A_d \mathbb{E}\big[x_k \mid y_1, \dots, y_k\big] + B_d \underbrace{\mathbb{E}\big[u_k \mid y_1, \dots, y_k\big]}_{u_k} + \underbrace{\mathbb{E}\big[v_k \mid y_1, \dots, y_k\big]}_{\mathbb{E}[v_k]=0} \\
&= A_d \mathbb{E}\big[x_k \mid y_1, \dots, y_k\big] + B_d u_k \\
&= A_d \hat{x}_k + B_d u_k \ ,
\end{aligned}$$

where we have used, on the one hand, that u_k only depends, by assumption, on $y_1, \dots, y_k$ (item 1 in Definition 7.10), and, on the other hand, that v_k is centered and independent of $y_1, \dots, y_k$ by (7.16).

We now evaluate the matrix L_{k+1} solution of $\mathsf{cov}(x_{k+1}, i_{k+1}) = L_{k+1}\mathsf{D}(i_{k+1})$. First, let us compute $\mathsf{D}(i_{k+1})$ thanks to

$$\begin{aligned}
i_{k+1} &= y_{k+1} - \mathbb{E}\big[y_{k+1} \mid y_1, \dots, y_k\big] \text{ by (7.20)} \\
&= C x_{k+1} + w_{k+1} - \mathbb{E}\big[C x_{k+1} + w_{k+1} \mid y_1, \dots, y_k\big] \text{ by (7.16)} \\
&= C x_{k+1} + w_{k+1} - C\mathbb{E}\big[x_{k+1} \mid y_1, \dots, y_k\big] - \underbrace{\mathbb{E}\big[w_{k+1} \mid y_1, \dots, y_k\big]}_{\mathbb{E}[w_{k+1}]=0} \\
&= C\left(x_{k+1} - \hat{x}_{k+1}^{-}\right) + w_{k+1} \ .
\end{aligned}$$

By (7.16), w_{k+1} is independent of $x_{k+1} - \hat{x}^-_{k+1}$, so that $\mathsf{D}(i_{k+1})$ is invertible. Indeed, we have that

$$\begin{aligned}\mathsf{D}(i_{k+1}) &= \mathsf{D}(C\left(x_{k+1} - \hat{x}^-_{k+1}\right)) + \mathsf{D}(w_{k+1}) \\ &= C\mathsf{D}(x_{k+1} - \hat{x}^-_{k+1})C^\top + R \geq R > 0\,,\end{aligned}$$

by assumption on the measurement noises. Then, we compute

$$\begin{aligned}\mathsf{cov}(x_{k+1}, i_{k+1}) &= \mathsf{cov}(x_{k+1} - \hat{x}^-_{k+1}, i_{k+1}) + \mathsf{cov}(x^-_{k+1}, i_{k+1}) \\ &= \mathsf{cov}(x_{k+1} - \hat{x}^-_{k+1}, i_{k+1}) \\ &\quad\text{by independence of } y_1, \ldots, y_k \text{ and } i_{k+1} \\ &= \mathsf{cov}(x_{k+1} - \hat{x}^-_{k+1}, C\left(x_{k+1} - \hat{x}^-_{k+1}\right) + w_{k+1}) \\ &\quad\text{by the expression of } i_{k+1} \text{ computed above} \\ &= \mathsf{D}(x_{k+1} - \hat{x}^-_{k+1})C^\top .\end{aligned}$$

With the notation (7.22), we have, up to now, obtained the expression

$$\hat{x}_{k+1} = A_d\hat{x}_k + B_d u_k + L_{k+1}\big(y_{k+1} - C(A_d\hat{x}_k + B_d u_k)\big)\ .$$

It remains to compute a recursive formula for the matrix L_{k+1} by means of

$$\begin{aligned}P^-_{k+1} &= \mathsf{D}(x_{k+1} - \hat{x}^-_{k+1}) \\ &= \mathsf{D}(A_d x_k + B_d u_k + v_k - A_d\hat{x}_k - B_d u_k) \\ &= \mathsf{D}\big(A_d(x_k - \hat{x}_k)\big) + \mathsf{D}(v_k) \\ &= A_d\mathsf{D}(x_k - \hat{x}_k)A_d^\top + Q\ .\end{aligned}$$

Regarding the dispersion $P_k = \mathsf{D}(x_k - \hat{x}_k)$ of the error (7.19), it satisfies:

$$\begin{aligned}P_{k+1} &= \mathsf{D}(x_{k+1} - \hat{x}_{k+1}) \\ &= \mathsf{D}(x_{k+1} - \hat{x}^-_{k+1} - L_{k+1}i_{k+1}) \\ &= \mathsf{D}(x_{k+1} - \hat{x}^-_{k+1}) \\ &\quad -\mathsf{cov}(x_{k+1} - \hat{x}^-_{k+1}, L_{k+1}i_{k+1})\mathsf{D}(L_{k+1}i_{k+1})^{-1}\mathsf{cov}(x_{k+1} - \hat{x}^-_{k+1}, L_{k+1}i_{k+1})^\top \\ &\quad\text{by Lemma C.13 (with } X = x_{k+1} - \hat{x}^-_{k+1} \text{ and } Y = L_{k+1}i_{k+1}) \\ &= P^-_{k+1} - L_{k+1}CP^-_{k+1}\ .\end{aligned}$$

This concludes the proof. □

Remark 7.14 If the matrices A_d, B_d, C, Q and R depend on time k, the previous result and formulas remain valid. This makes it possible to compute nonstationary optimal state estimators (in the sense of conditional expectation). ◇

7.3.2 Convergence of the Filter. Links with Observability

Let us write the recursive equations (7.23) of the Kalman-Bucy filter, but skipping the intermediary variables $\hat{x}_k^-$ and P_k:

$$\begin{cases} \hat{x}_{k+1} = A_d\hat{x}_k + (B_d - L_{k+1}CB_d)u_k + L_{k+1}(y_{k+1} - CA_d\hat{x}_k) \\ P_{k+1}^- = Q + A_dP_k^-A_d^\top - A_dP_k^-C^\top(R + CP_k^-C^\top)^{-1}CP_k^-A_d^\top \\ L_{k+1} = P_{k+1}^-C^\top(CP_{k+1}^-C^\top + R)^{-1}. \end{cases}$$

Under this last form, the Kalman-Bucy filter (7.23) appears like an observer (discussed in § 6.6) of equation

$$\hat{x}_{k+1} = A_d\hat{x}_k + (B_d - L_{k+1}CB_d)u_k + L_{k+1}(y_{k+1} - CA_d\hat{x}_k)\ , \tag{7.24}$$

with time varying gain matrix L_k given by the induction

$$\begin{cases} P_{k+1}^- = Q + A_dP_k^-A_d^\top - A_dP_k^-C^\top(R + CP_k^-C^\top)^{-1}CP_k^-A_d^\top \\ L_{k+1} = P_{k+1}^-C^\top(CP_{k+1}^-C^\top + R)^{-1}. \end{cases}$$

Indeed, it suffices to compare (7.24) to the formula (6.36) of the observer in Theorem 6.28 (but not the one to be found in Theorem 6.27).

As discussed in § 6.6, an observer has the property that the discrepancy between the estimate of the state and the state itself converges towards 0. Here, due to random terms, such a notion of convergence is not adapted. However, we show that, under proper assumptions, the estimation error (7.19) of the Kalman-Bucy filter converges, *in the sense of convergence in distribution*, towards a centered Gaussian distribution, and the time-varying gain converges towards a fixed stabilizing gain.

Proposition 7.15 *Suppose that*

1. *the couple (A_d, C) (or (A_d, CA_d)) is observable;*
2. *there exists a square root G of the matrix Q ($Q = GG^\top$) such that the couple (A_d, G) is controllable.*

Then, the matrices P_k^-, P_k and L_k of the Kalman-Bucy filter (7.23) *converge towards matrices denoted respectively P^-, P and L, that do not depend on the initial conditions $\overline{x}_0$ and P_0. The matrix $A_d - LCA_d$ is asymptotically stable and the estimation error* (7.19) *converges in distribution towards a centered Gaussian distribution with dispersion matrix P.*

Moreover, the matrix P^- is positive definite *and is obtained as limit of any sequence*

$$P_{k+1}^- = Q + A_dP_k^-A_d^\top - A_dP_k^-C^\top(R + CP_k^-C^\top)^{-1}CP_k^-A_d^\top\ , \tag{7.25}$$

for any initial condition $P_0^- \geq 0$. *We also have that*

$$L = P^- C^\top (R + CP^- C^\top)^{-1} \text{ and } P = (I - LC)P^- .$$

Proof First, let us apply Proposition A.4 to the couple $(A_d^\top, C^\top)$ and to the couple $(A_d^\top, G^\top)$. This assures the existence of a positive symmetric matrix P^-, stationary solution of the Riccati equation (7.25).

Regarding the error e_k in (7.19), one observes by (7.24) and (7.16) that is satisfies the induction

$$\begin{aligned} e_{k+1} &= A_d e_k + L_{k+1} C B_d u_k - L_{k+1}(C x_{k+1} + w_{k+1} - C A_d \hat{x}_k) \\ &= A_d e_k + L_{k+1} C B_d u_k - L_{k+1}(C A_d x_k + C B_d u_k + C v_k + w_{k+1} - C A_d \hat{x}_k) \\ &= A_d e_k - L_{k+1}(C A_d e_k + C v_k + w_{k+1}) \\ &= (A_d - L_{k+1} C A_d) e_k - L_{k+1}(C v_k + w_{k+1}) \,. \end{aligned}$$

Notice that, by the assumptions of Theorem 7.12, the sequence $(Cv_k + w_{k+1})_{k\in\mathbb{N}}$ is made of independent random variables. Moreover, $Cv_k + w_{k+1}$ is Gaussian as a linear combination of independent Gaussian random variables, by Proposition C.9 and Proposition C.7. Thus, $(Cv_k + w_{k+1})_{k\in\mathbb{N}}$ is a Gaussian white noise. Thus, the sequence $e_1, \ldots, e_k$ of errors is generated by a linear dynamical system excited by a Gaussian white noise, with an independent initial condition $e_0 = x_0 - \overline{x}_0$. Therefore, by Proposition C.7 and Proposition C.8, the random variables e_k are Gaussian with zero mean. By Definition C.5 and Proposition C.6, the characteristic function Φ_{e_k} of the error e_k is thus given by:

$$\Phi_{e_k}(\theta) = \exp(-\frac{1}{2}\theta^\top \mathsf{D}(e_k)\theta) \,.$$

Now, $\mathsf{D}(e_k) = P_k$ converges towards $P = (I - LC)P^-$, so that $\Phi_{e_k}(\theta)$ converges towards $\exp(-\frac{1}{2}\theta^\top P\theta)$. By Definition C.5, this last term is the characteristic function of a Gaussian centered distribution with dispersion matrix P. We conclude because simple convergence of characteristic functions implies convergence in distribution. □

7.4 Formulas in the Continuous-Time Case

We now enounce the continuous-time counterparts of the results of § 7.2 and 7.3.

▷ We refer the reader to [12,33].

7.4.1 Optimization in Finite Horizon

Consider the controlled linear dynamical system in continuous-time

$$\dot{x} = Ax + Bu \ , \tag{7.26}$$

and, for $T > 0$, the criterion

$$J(u) = \int_0^T \Big(u(s)^\top Ru(s) + x(s)^\top Qx(s) \Big) \, \mathrm{d}s + x(T)^\top Q_T x(T) \ , \tag{7.27}$$

where Q_T is a symmetric positive matrix of size n. Here, in accordance with Remarks 2.4 and 3.1, the notations x and u correspond to *continuous trajectories* on the half-line, that is, elements of $C^0(\mathbb{R}_+, \mathbb{R}^n)$ and $C^0(\mathbb{R}_+, \mathbb{R}^m)$, respectively.

Proposition 7.16 *If* $S_0 = S(0)$ *is the terminal value of the* Riccati backward differential equation

$$\dot{S}(t) + A^\top S(t) + S(t)A - S(t)BR^{-1}B^\top S(t) + Q = 0 \ , \ \ S(T) = Q_T, \tag{7.28}$$

the minimum of the criterion (7.2) *is*

$$\min_u J(u) = x_0^\top S_0 x_0 \ .$$

Moreover, this minimum is achieved for the following linear state feedback control

$$u^\star(t) = -R^{-1}B^\top S(t)x^\star(t) \ ,$$

where $x^\star(t)$ is the optimal trajectory corresponding to the application of the previous optimal control, namely, $x_0^\star = x_0$ and

$$\dot{x}^\star = Ax^\star + Bu^\star = \big(A - BR^{-1}B^\top S(t)\big)x^\star.$$

Proof The proof is inspired from [12].

By Exercise 7.6.3 (changing t in $T - t$), we have that the matrix $S(t)$ is defined for all $t \leq T$. Moreover, we can write

$$\begin{aligned} &x(t)^\top S(T)x(t) - x(0)^\top S(0)x(0) = \\ &\quad \int_0^T \frac{d}{dt}\big(x(t)^\top S(t)x(t)\big) \, \mathrm{d}t = \int_0^T \Big(x(t)^\top \dot{S}(t)x(t) + \big(Ax(t) + Bu(t)\big)^\top S(t)x(t) \\ &\qquad\qquad + x(t)^\top S(t)\big(Ax(t) + Bu(t)\big) \Big) \, \mathrm{d}t \ . \end{aligned}$$

As $S(T) = Q_T$ in (7.28), we deduce by (7.27) that

$$\begin{aligned} J(u) &= x_0^\top S(0)x_0 \\ &\quad + \int_0^T (u(t)^\top; x(t)^\top) \begin{pmatrix} R & B^\top S(t) \\ S(t)B & \dot{S}(t) + A^\top S(t) + S(t)A + Q \end{pmatrix} \begin{pmatrix} u(t) \\ x(t) \end{pmatrix} \mathrm{d}t \\ &= x_0^\top S(0)x_0 \\ &\quad + \int_0^T (u(t)^\top; x(t)^\top) \begin{pmatrix} R & B^\top S(t) \\ S(t)B & S(t)BR^{-1}B^\top S(t) \end{pmatrix} \begin{pmatrix} u(t) \\ x(t) \end{pmatrix} \mathrm{d}t \\ &= x_0^\top S(0)x_0 + \int_0^T \big(u(t) + R^{-1}B^\top S(t)x(t)\big)^\top R\big(u(t) + R^{-1}B^\top S(t)x(t)\big)\,\mathrm{d}t . \end{aligned}$$

This ends the proof. □

Remark 7.17 If the matrices A, B, Q and R depend (continuously) on the time t, the previous result and formulas remain valid. ◇

7.4.2 Optimization in Infinite Horizon. Links with Controllability

We now show that the linear control with stationary gains discussed in Chap. 5 minimizes an intertemporal criterion.

Let $\mathcal{U} \subset C^0(\mathbb{R}_+, \mathbb{R}^m)$ be the set of control trajectories u such that the state trajectory x given by (7.26) converges towards 0 and that

$$\int_0^{+\infty} \Big(u(s)^\top Ru(s) + x(s)^\top Qx(s)\Big)\,\mathrm{d}s < +\infty .$$

We define an intertemporal criterion J on $\mathcal{U}$ by

$$J(u) = \int_0^{+\infty} \Big(u(s)^\top Ru(s) + x(s)^\top Qx(s)\Big)\,\mathrm{d}s , \quad \forall u \in \mathcal{U} . \tag{7.29}$$

The proof of Proposition 7.16 can easily be adapted to establish the following result.

Proposition 7.18 *Suppose that there exists a symmetric positive matrix S such that*

1. *the matrix S is solution of the* algebraic (or stationary) continuous-time Riccati equation
$$A^\top S + SA - SBR^{-1}B^\top S + Q = 0 \;; \tag{7.30}$$
2. *the square matrix $A - BR^{-1}B^\top S$ is asymptotically stable.*

Then, the minimum of the criterion (7.29) *over the set* $\mathcal{U}$ *of control trajectories is given by*

$$\min_{u \in \mathcal{U}} J(u) = x_0^\top S x_0 \; . \tag{7.31}$$

Moreover, this minimum is achieved for the following linear state feedback control

$$u^\star(t) = -R^{-1} B^\top S x^\star(t) \; , \tag{7.32}$$

where $x^\star(t)$ *is the optimal trajectory corresponding to the application of the previous optimal control, namely* $x_0^\star = x_0$ *and*

$$\dot{x}^\star = A x^\star + B u^\star = (A - BR^{-1}B^\top S) x^\star. \tag{7.33}$$

The existence of a matrix S that satisfies the two assumptions of Proposition 7.18 is assured under the following assumptions on the matrices A of dynamics, B of control, and Q of state ponderation.

Proposition 7.19 *Suppose that*

1. *the couple* (A, B) *is controllable;*
2. *there exists a square root* H *of the matrix* Q $(Q = H^\top H)$ *such that the couple* (A, H) *is observable.*

Then, there exists a unique symmetric positive matrix S *which satisfies the two assumptions of Proposition 7.18. Moreover, the matrix* S *is* positive definite *and is obtained as the limit, for* $t \to +\infty$, *of any solution of the differential equation*

$$-\dot{S}(t) + A^\top S(t) + S(t)A - S(t)BR^{-1}B^\top S(t) + Q = 0 \tag{7.34}$$

for any initial condition $S(0) \geq 0$.

Proof The uniqueness of the symmetric positive matrix S solution of (7.30) comes from Proposition 7.18, since S is associated with the minimum of the criterion (7.29). The existence of the matrix S is shown in Exercise 7.6.4. The rest of the proof is inspired from [66].

Consider Eq. (7.30) where A is replaced by $-A$. There exists a unique symmetric matrix S_- such that $-S_-$ is positive, and that

$$\begin{aligned} 0 &= A^\top S_- + S_- A - S_- BR^{-1}B^\top S_- + Q, \; S_- < 0 \\ -A_- &= -(A - BR^{-1}B^\top S_-) \text{ is asymptotically stable.} \end{aligned}$$

Denoting $S_+ = S$ the unique symmetric positive matrix solution of (7.30), we have that:

$$0 = A^\top S_+ + S_+ A - S_+ B R^{-1} B^\top S_+ + Q \;,\; S_+ > 0$$
$$A_+ = A - B R^{-1} B^\top S_+ \text{ is asymptotically stable.}$$

By using Corollary 5.17, we first show that the couple (A_-, B) is controllable. Indeed, suppose there exists a vector x such that

$$x^\top B = x^\top A_- B = \cdots = x^\top A_-^{n-1} B = 0 \;,$$

then, as $A_- = A - BR^{-1}B^\top S_-$, we deduce that $x^\top A_- B = x^\top AB$ because $x^\top B = 0$ and, step by step, that $x^\top A_-^k B = x^\top A^k B$. Thus, we obtain

$$x^\top B = x^\top AB = \cdots = x^\top A^{n-1} B = 0 \;,$$

and we conclude that $x = 0$ by controllability of the couple (A, B) and by Corollary 5.17.

Now, we show that $S(t) \to_{t\to+\infty} S_+$. Setting $\Delta(t) = S(t) - S_-$, the computation shows that

$$-\dot{\Delta}(t) + A_-^\top \Delta(t) + \Delta(t) A_- - \Delta(t) B R^{-1} B^\top \Delta(t) = 0 \;,$$

with $\Delta(0) = S(0) - S_- \geq -S_- > 0$ since $S(0) \geq 0$. Let us introduce

$$\Gamma(t) = e^{-tA_-^\top} \left(\Delta(0)^{-1} + \int_0^t e^{sA_-} B R^{-1} B^\top e^{sA_-^\top} \, ds \right) e^{-tA_-} \;,$$

solution of the linear matrix equation

$$-\dot{\Gamma}(t) + A_-^\top \Gamma(t) + \Gamma(t) A_- - B R^{-1} B^\top = 0 \;,\; \Gamma(0) = \Delta(0)^{-1} \;.$$

The matrix $\Gamma(t)$ is invertible for all t thanks to Lemma 5.16, since the couple (A_-, B) is controllable. As $\Gamma(t)^{-1}$ satisfies the same equation as $\Delta(t)$ with the same initial condition, we have the equality $\Gamma(t)^{-1} = \Delta(t)$, and thus $\Delta(t)$ is invertible.

In the same way, we can show that $\Delta_\infty = S_+ - S_-$ is invertible, with inverse Γ_∞ solution of:

$$-A_-^\top \Gamma_\infty - \Gamma_\infty A_- + B R^{-1} B^\top = 0 \;.$$

By substraction of the last two equations, we obtain

$$\frac{d}{dt}(\Gamma(t) - \Gamma_\infty) = -A_-^\top (\Gamma(t) - \Gamma_\infty) - (\Gamma(t) - \Gamma_\infty) A_- \;.$$

As $-A_-$ is asymptotically stable, we conclude that $\Gamma(t) \to_{t\to+\infty} \Gamma_\infty$. By inversion, this gives $\Delta(t) \to_{t\to+\infty} \Delta_\infty$, namely $S(t) - S_- \to_{t\to+\infty} S_+ - S_-$ and thus $S(t) \to_{t\to+\infty} S_+$. This ends the proof since $S_+ = S$. □

7.4.3 Asymptotic Observer

The following proposition is proven along the same lines as Proposition 7.19.

Proposition 7.20 *Suppose that*

1. *the couple (A, C) is observable;*
2. *there exists a square root G of the matrix Q ($Q = GG^\top$) such that the couple (A, G) is controllable.*

Then, there exists a unique symmetric positive matrix S solution of the algebraic Riccati equation

$$AS + SA^\top - SC^\top R^{-1}CS + Q = 0 \tag{7.35}$$

such that $A - SC^\top R^{-1}C$ is asymptotically stable. Moreover, this matrix S is positive definite *and is obtained as the limit of any solution of the differential equation*

$$-\dot{S}(t) + AS(t) + S(t)A^\top - S(t)C^\top R^{-1}CS(t) + Q = 0 \tag{7.36}$$

for any initial condition $S(0) \geq 0$.

7.5 Practical Set up

As discussed in § 7.2.3, the methods developed in this chapter put the problem of the *choice* of the closed-loop system modes onto the *choice* of the ponderation matrices Q and R in an intertemporal criterion.

For the regulator, these matrices Q and R, respectively, weigh the state and the control in the criterion (7.8).

For the observer, or for the Kalman-Bucy filter, these matrices Q and R are the respective dispersion matrices of state and of measurement noises (assumed white Gaussian). The knowledge of the precision of the measurement sensors makes it possible to quantify R: each of the diagonal terms is of the order of the square of the standard deviation on the measurement (the higher, the less precise). Regarding the matrix Q, each of the diagonal terms is of the order of the square of the standard deviation of perturbations affecting the dynamics.

7.6 Exercises

Exercise 7.6.1 *(Kalman-Bucy filter with correlated noises)*

Consider the controlled and observed linear dynamical system with additive noises (7.16), where we suppose that

- the control u_k depends *affinely* on the observations $y_1, \ldots, y_k$ up to time k;

- the initial state x_0 is a Gaussian random vector;
- $(v_k, w_k)_{k\geq 1}$ is a stationary Gaussian white noise with covariance matrix

$$\mathsf{D}\begin{pmatrix} v_k \\ w_k \end{pmatrix} = \begin{pmatrix} Q & S \\ S^\top & R \end{pmatrix};$$

- x_0 and $(v_k, w_k)_{k\in\mathbb{N}}$ are independent.

Such a system is called *Gaussian linear dynamical system with correlated noises.* The two last assumptions above replace the last three of Definition 7.10.

1. Using $y_k = Cx_k + w_k$ in (7.16), write (7.16) under the form

$$\begin{cases} x_{k+1} = A_d x_k + B_d u_k + v_k + T(y_k - Cx_k - w_k)\,, & k \geq 0 \\ y_k = Cx_k + w_k\,, & k \geq 1\,, \end{cases}$$

   where T is any matrix of appropriate dimensions. Deduce that the model above can be put under the form (7.16) with a new matrix of dynamics, a new control, and a new state noise.
2. Show that, if $T = SR^{-1}$, the new state noise and the measurement noises are independent white noises.
3. Deduce that, with such a choice of matrix T, the assumptions of Proposition 7.11 are satisfied.
4. Give the equations of the filter in the case of correlated noises.

Exercise 7.6.2 *(Estimation of a measurement bias)*
Consider a sensor, that displays an unknown bias. We model this under the form

$$\begin{cases} x_{k+1} = A_d x_k + B_d u_k + v_k, k \geq 0 \\ y_k = Cx_k + b + w_k, k\geq 1\,, \end{cases}$$

where b is an unknown scalar representing this bias.

1. We set $b_k = b$. Show that the extended vector $z_k = \begin{pmatrix} x_k \\ b_k \end{pmatrix}$ satisfies the dynamical equation

$$\begin{cases} z_{k+1} = \widetilde{A}_d z_k + \widetilde{B}_d u_k + \widetilde{v}_k, k \geq 0 \\ y_k = \widetilde{C} z_k + w_k, k \geq 1 \end{cases}$$

   with

$$\widetilde{A}_d = \begin{pmatrix} A_d & 0 \\ 0 & 1 \end{pmatrix}, \widetilde{B}_d = \begin{pmatrix} B_d \\ 0 \end{pmatrix}, \widetilde{v}_k = \begin{pmatrix} v_k \\ 0 \end{pmatrix}, \widetilde{C} = (C \quad 1)\,.$$

2. Provide the recursive formulas giving the estimates $\hat{x}_k$ and $\hat{b}_k$.
3. Explain why the definition of $\widetilde{v}_k$ prevents us from applying Proposition 7.15 and from obtaining asymptotic formulas. In practice, for stability issues, we model $\widetilde{v}_k$ as a noise with nondegenerate dispersion matrix.

Exercise 7.6.3 *(Riccati equation in continuous-time)*

We show that the *Riccati differential equation*

$$-\dot{\Pi}(t) + F^\top \Pi(t) + \Pi(t)F - \Pi(t)GR^{-1}G^\top \Pi(t) + Q = 0 \tag{7.37}$$

admits a solution for all time $t \geq 0$ as soon as $R > 0$, $Q \geq 0$ and $\Pi(0) \geq 0$.

1. Let $\Pi(0) \geq 0$ be given. Consider $T > 0$ such that Eq. (7.37) admits a solution on $[0, T]$ by the Cauchy-Lipschitz Theorem. Moreover, let the vector $z(t)$ be the solution of the following linear differential equation

$$\dot{z}(t) = \big(-F + GR^{-1}G^\top \Pi(t)\big)z(t)\ ,\quad 0 \leq t \leq T\ ,\quad z(t) = z_0\ .$$

   Show that $t \mapsto z(t)^\top \Pi(t)z(t)$ is an increasing function. Deduce that $z(t)^\top \Pi(t)$ $z(t) \geq 0$ for all $t \in [0, T]$ and, by properly selecting z_0 for a fixed t, that $\Pi(t) \geq 0$.
2. Let $P(t)$ be the solution of the matrix linear differential equation

$$-\dot{P}(t) + F^\top P(t) + PF(t) + Q = 0,\ P(0) = \Pi(0)\ ,\quad 0 \leq t\ ,$$

   that is, for all t,

$$P(t) = e^{tF^\top}\left(\Pi(0) + \int_0^t e^{-sF^\top} Qe^{-sF}\,\mathrm{d}s\right)e^{tF}.$$

   Check that $\Delta(t) = P(t) - \Pi(t)$ satisfies

$$-\dot{\Delta}(t) + F^\top \Delta(t) + \Delta(t)F = -\Pi(t)GR^{-1}G^\top \Pi(t) \leq 0\ ,\quad 0 \leq t \leq T.$$

   Deduce that $\Delta(t) \geq 0$ by evaluating $z(t)^\top \Delta(t)z(t)$ where $z(t)$ solves $\dot{z}(t) = -Fz(t)$.
3. Observe that the inequality $0 \leq \Pi(t) \leq P(t)$ implies that $\Pi(t)$ is a priori bounded above on all bounded time interval, hence is defined for all $t \geq 0$. Conclude.

Exercise 7.6.4 *(Stationary Riccati equation in continuous-time)*

We are going to show that the *stationary Riccati equation in continuous-time*

$$F^\top \Pi + \Pi F - \Pi GR^{-1}G^\top \Pi + Q = 0 \tag{7.38}$$

admits a solution such that $F - GR^{-1}G^\top \Pi$ is asymptotically stable as soon as the couple (F, G) is controllable and that there exists a square root H of the matrix Q ($Q = H^\top H$) such that the couple (F, H) is observable. We refer the reader to Appendix A.

1. If $(\Pi_0(t),\ t \geq 0)$ is the solution of (7.37) for $\Pi_0(0) = 0$, check, thanks to Proposition 7.16, that

$$x_0^\top \Pi_0(t) x_0 = \min_u \int_0^t \left(u(s)^\top R u(s) + x(s)^\top Q x(s)\right) \mathrm{d}s \ ,$$

where $\dot{x}(t) = Fx(t) + Gu(t)$. Deduce that $t \mapsto x_0^\top \Pi_0(t) x_0$ is an increasing function.

2. Show that, thanks to the controllability of the couple (F, G), this last function is bounded above. Deduce that $x_0^\top \Pi_0(t) x_0$ converges towards a limit, denoted $\varphi(x_0)$.
3. Show that $\varphi(x_0)$ can be expressed under the form $\varphi(x_0) = x_0^\top \Pi_\infty x_0$.
4. Deduce that $\Pi_0(t) \to \Pi_\infty$ and, by (7.37), that $\dot{\Pi}_0(t)$ also admits a limit that cannot be zero. Thus, Π_∞ satisfies (7.38).
5. Show that

$$\begin{aligned} x_0^\top \Pi_\infty x_0 = 0 &\iff \min_u \int_0^t \Big(u(s)^\top R u(s) + x(s)^\top Q x(s)\Big)\, \mathrm{d}s = 0 \ , \quad \forall t \geq 0 \\ &\text{for } \dot{x}(t) = Fx(t) + Gu(t) \ , \quad x(0) = x_0 \\ &\iff G^\top \Pi_0(t) x(t) = 0 \text{ and } Hx(t) = 0 \ , \quad \forall t \geq 0 \\ &\text{for } \dot{x}(t) = \big(F - GR^{-1}G^\top \Pi_0(t)\big) x(t) \ , \quad x(0) = x_0 \ . \end{aligned}$$

Deduce that $x_0 = 0$ by observability of the couple (F, H). Thus, Π_∞ is definite positive.

6. Check that (7.38) can be written as

$$(F - GR^{-1}G^\top \Pi)^\top \Pi + \Pi(F - GR^{-1}G^\top \Pi) + \Pi G R^{-1} G^\top \Pi + Q = 0 \ .$$

Show that the function $\mathfrak{V}(x) = x^\top \Pi_\infty x$ satisfies the assumptions of the LaSalle Theorem 4.31 for the system

$$\dot{x} = (F - GR^{-1}G^\top \Pi_\infty) x \ .$$

Deduce that the solution $x(t)$ converges towards the largest invariant subset $\mathbb{I}$ of $\{x \mid G^\top \Pi_\infty x = 0 \text{ and } Hx = 0\}$. In the same way as in the previous item, show that $\mathbb{I}$ is reduced to the singleton $\{0\}$ thanks to the observability of the couple (F, H). Deduce that the matrix $F - GR^{-1}G^\top \Pi_\infty$ is asymptotically stable.

Part III
Disturbance Rejection and Polynomial Approach

Chapter 8
Polynomial Representation

8.1 Introduction

We have discussed how to stabilize a linear controllable and observable system given, in Chap. 5, by a continuous-time input-output representation

$$\begin{cases} \dot{x} = Ax + Bu \\ y = Cx \end{cases}$$

or, in Chap. 6, by a discrete-time input-output representation

$$\begin{cases} x_{k+1} = A_d x_k + B_d u_k \\ y_k = C x_k \end{cases}$$

using, for example, a quadratic optimization approach as developed in Chap. 7, with u denoting the m-input or control vector, y the p-output vector and x the n-state vector.

The separation principle between estimation and control, discussed in § 5.6 and 6.7, makes it possible to determine a stabilizing control law by fixing $n(m+p)$ parameters, nm for the gain regulator matrix and np for the gain observer matrix. But, to stabilize the closed-loop system, it is sufficient to place $2n$ poles, n "regulator poles" of the matrix $A - BK$, or of the matrix $A_d - B_d K$ in discrete-time, and n "observer poles" of the matrix $A - LC$, or of the matrix $A_d - LC$. Therefore, in the multivariable case, we still have

$$l = n(m + p - 2) \tag{8.1}$$

degrees of freedom, l being strictly positive since $m + p > 2$. Therefore, it is interesting to use these degrees of freedom to obtain a desired input-output behavior, precisely in the case of perturbations (neglected dynamics, noises of sensors or actuators, etc.), and then to bring robustness to the closed-loop system. This is

B. d'Andréa-Novel and M. De Lara, *Control Theory for Engineers*,
DOI: 10.1007/978-3-642-34324-7_8,

possible by acting on the *zeros* of the system, the poles having already been fixed, for example, by an observer-regulator quadratic synthesis as highlighted in Chap. 7. In fact, we have already introduced in § 3.5 the definition and properties of the zeros of a transfer function, relative to the disturbance rejection problem. For example, we have seen in Proposition 3.32 that, if the transfer function of an l.c.s. system has a zero at $s = 0$, constant bias on the input is asymptotically rejected.

Moreover, it is emphasized that the so-called *polynomial representation* is well adapted to the computation of closed-loop transfer functions between disturbances and outputs of the system. In the multivariable case, it remains to parameterize the controller using the degrees of freedom and to fix them, for example, to place some zeros at $s = 0$ in the closed-loop transfer functions between some disturbances and outputs. This makes it possible to asymptotically reject the effect of constant perturbations on the system. It must be noticed that this rejection is realized without needing to build an asymptotic observer for each observable bias, and consequently without extending the size of the controller, as it is often the case in a classical state-space context (see Exercise 7.6.2 and the following Example 8.1).

Example 8.1 Consider the inverted pendulum fixed on a cart moving on an horizontal bench introduced in § 2.3.1. The matrices A, B and C, of the tangent linear system in the neighborhood of the unstable equilibrium are given in (5.4). Suppose that there is a constant bias b_θ on the measurement of the angle θ (see Fig. 2.3). The observations in Example 5.3 are now given by:

$$\begin{cases} y_1 = z \\ y_2 = \theta + b_\theta \ . \end{cases}$$

If this bias b_θ is neglected, when applying the method of the observer-regulator described in § 5.6, denoting $\hat{x}$ an asymptotic observer of the state $x = (z, \theta, \dot{z}, \dot{\theta})^\top$, it can be easily shown that the asymptotic bias on the observer is of the form

$$x_\infty - \hat{x}_\infty = (A - LC)^{-1} L b_o$$

where $b_o = (0 \quad b_\theta)^\top$ denotes the vector of measurements bias and $L_{4\times 2}$ the observer gain matrix (see § 5.10.2 for the method).

Then, applying the linear state feedback $u = -K\hat{x}$ (instead of $u = -Kx$) introduces an asymptotic bias b_u on the control given by

$$b_u = -K(A - LC)^{-1} L b_o \ ,$$

where $K_{1\times 4}$ denotes the regulator gain matrix. Therefore, the existence of the bias b_u on the control induces an asymptotic bias z_∞ on the position z, given by

$$z_\infty = \frac{b_u}{K_{11}} \ ,$$

which means that the asymptotic position of the cart is biased. Let us point out that the position of the pendulum is not biased (we have $\theta_\infty = 0$), which is natural since, without pertubation on the dynamics, the only possible equilibrium for the pendulum is the vertical position.

To eliminate the bias z_∞ on the position z of the cart, we can introduce an observer for b_θ and then increase the size of the state vector by adding the variable b_θ assumed to be constant, namely with the dynamics $\dot{b}_\theta = 0$. This leads to the augmented system

$$\begin{cases} \dot{\widetilde{x}}_a = A_a \widetilde{x}_a + B_a u \\ y = C_a \widetilde{x}_a \ , \end{cases}$$

with the augmented state

$$\widetilde{x}_a = \left(b_\theta, z, \theta, \dot{z}, \dot{\theta}\right)^\top$$

and

$$A_a = \begin{pmatrix} 0\,0\,0 & 0\,0 \\ 0 & \\ 0 & A \\ 0 & \\ 0 & \end{pmatrix}, \quad B_a = \begin{pmatrix} 0 \\ B \end{pmatrix}.$$

If K is a gain matrix stabilizing $A - BK$, then, due to the form of the matrices A_a and B_a, the same gain matrix K gives, for the augmented system,

$$\widetilde{x}_a \to 0 \quad \text{and} \quad b_\theta = \text{constant}.$$

The bias b_θ is not controllable, but it can be easily shown that it is observable. An asymptotic observer of the augmented system can then be built, and one can find a suitable gain matrix L_a of dimension 5×2 such that the matrix $(A_a - L_a C_a)$ is asymptotically stable. Applying the control

$$u = -K_a \hat{\widetilde{x}}_a = -\left[\, 0 \ K \,\right] \hat{\widetilde{x}}_a$$

leads, for the closed-loop system with state $\widetilde{x}_a$ and $e_a = \widetilde{x}_a - \hat{\widetilde{x}}_a$, to the dynamics:

$$\begin{cases} \dot{\widetilde{x}}_a = (A_a - B_a K_a)\widetilde{x}_a + B_a K_a e_a \\ \dot{e}_a = (A_a - L_a C_a) e_a \ . \end{cases}$$

It is straightforward to observe that $e_a(t) \to 0$, which means that the bias on the observer, and therefore on the asymptotic value of the cart's position, has been eliminated.

It can be noticed that the estimate $\hat{b}_\theta$ of the bias does not directly appear in the expression of the control, but that it modifies the observer's dynamics $\hat{\widetilde{x}}_a$ and that the control is proportional to $\hat{\widetilde{x}}_a$. △

As illustrated above, in a state-space approach, one can construct specific observers to asymptotically reject observable bias, but this method increases the size of the state and therefore the complexity of the control law. To obtain a minimal size controller, one could use geometrical techniques developed by Wonham [70] and Willems [67], leading to some conditions for global disturbance rejection on the output. However, even if the global disturbance rejection problem has no solution, one could try, at least, to eliminate disturbances on some components of the output.

For linear systems, the state-space representation and the polynomial representation are conceptually equivalent, but the classical state-space representation does not make it possible to access the partial transfer functions between inputs (control variables or disturbances) and different components of the output. This is the reason why we use representations that make it easy to compute the different transfer functions connecting inputs and outputs, called *polynomial representations*. Then, it is interesting to use the degrees of freedom in the control to place zeros in suitable transfer functions between some disturbances and some outputs. This makes it possible to reject these disturbances with a *minimal size* controller.

Definitions are provided in § 8.2, and basic results on polynomial matrices in § 8.3. Poles and zeros, as well as stability, are discussed in § 8.4. Equivalence between linear differential systems is defined and characterized in § 8.5. Controllability and observability notions in the context of polynomial representation are introduced in § 8.6. In § 8.7, we establish the links between the state-space and the polynomial representations and we give a systematic way to compute the polynomial controller and observer forms from the classical state-space representation. In § 8.8, the closed-loop transfer functions from the input and the disturbances to the outputs are computed and, in § 8.9, we elaborate the parameterization of the controller with respect to the degrees of freedom in order to place some zeros when poles have been placed. Finally, an application to the inverted pendulum is provided in § 8.10.

- For more details on the theory of polynomial matrices, we refer the reader to [34–36].

8.2 Definitions

The use of polynomial matrices is an efficient way to represent systems of linear ordinary differential equations or difference equations in the discrete-time case, using the Laplace transform for continuous-time l.c.s. systems or the z-transform for discrete-time l.c.s. systems, as exposed in Appendix B.

Remark 8.2 In what follows, the complex Laplace variable s is also interpreted as the time differentiation operator $s = d/dt$ for continuous-time l.c.s. systems (see the differentiation Theorem L4 in B.1, [53]). In the same way, for discrete-time l.c.s. systems, the complex variable z is also identified with the advance operator (see the advance Theorem Z4 in § B.2). ◇

Let us introduce some definitions.

Definition 8.3 *A* polynomial matrix *(in s) is a matrix, each element of which is a polynomial in the complex variable s, with real coefficients.*

Remark 8.4 The variable s is often omitted, so that $P(s)$, or more simply P, represents a polynomial matrix. ◇

Definition 8.5 *an l.c.s. dynamical system with input u (of dimension m) and output y (of dimension p) is said to be in* polynomial form *if and only if there exist*

- *a vector ξ called* partial state *of dimension $\overline{n}$;*
- *polynomial matrices $P(s)_{\overline{n}\times\overline{n}}$, $Q(s)_{\overline{n}\times m}$ and $R(s)_{p\times\overline{n}}$,*

such that, in the frequency domain, the input-output relation is of the form:

$$\begin{cases} P(s)\xi = Q(s)u \\ \quad y = R(s)\xi \ . \end{cases} \tag{8.2}$$

Remark 8.6 Here, following Remark 8.2, the complex Laplace variable s is to be understood as the time differentiation operator $s = d/dt$. Also, in accordance with Remarks 2.4 and 3.1, the notations ξ, u, and y correspond to *smooth trajectories* on the half-line, that is, elements of $C^\infty(\mathbb{R}_+, \mathbb{R}^{\overline{n}})$, $C^\infty(\mathbb{R}_+, \mathbb{R}^m)$ and $C^\infty(\mathbb{R}_+, \mathbb{R}^m)$, respectively. ◇

Example 8.7 In classical mechanics, as discussed in § 1.3.3, a material point with mass m, position z and submitted to a force F, satisfies the differential equation:

$$\ddot{z} = F/m \ . \tag{8.3}$$

If the position z is assumed to be measured, a polynomial representation with partial state ξ_1 can be written directly as

$$\begin{cases} s^2\xi_1 = F/m \\ \quad y = \xi_1 \ . \end{cases} \tag{8.4}$$

△

Remarks

- Considering the previous example, it can be pointed out that (8.2) is a concise way to write linear differential equations in continuous-time.
- All the results obtained in this chapter can be easily extended to systems (8.2) having a direct link between the input and the output, namely

  $$y = R(s)\xi + W(s)u \ ,$$

  where $W(s)$ is a polynomial $p \times m$ matrix.

- The variable s denotes the Laplace variable corresponding to a continuous-time system but could also represent the advance or shift operator z for discrete-time systems. Therefore, all the results presented in this chapter could be directly extended to discrete-time systems.

To compute the transfer matrix of the l.c.s. system (8.2), the Laplace transform is applied to the differential equations (8.2). If $U(s)$, $Y(s)$ and $\Xi(s)$ denote respectively the Laplace transforms of $u(t)$, $y(t)$ and $\xi(t)$ (if they exist), we obtain:

$$\begin{cases} P(s)\Xi(s) = Q(s)U(s) \\ \quad Y(s) = R(s)\Xi(s) \; . \end{cases} \tag{8.5}$$

These equalities make it possible to define the *transfer matrix* of the l.c.s. system (8.2) by

$$H(s) = R(s)P^{-1}(s)Q(s) \; . \tag{8.6}$$

In a classical state-space representation, as explained in Chap. 5, two equivalent systems have a state vector with same dimension n. This is no more the case for polynomial representations, and this is why the notion of *partial state* of dimension $\overline{n} \leq n$ has been introduced in Definition 8.5.

Example 8.8 In the Example 8.7 of the material point in classical mechanics, the partial state ξ_1 is the scalar position z. But, it is clear that the velocity $\dot{\xi}_1$ is also needed to have a complete representation of the point's dynamics. The following polynomial representation can then be introduced

$$\begin{cases} \begin{pmatrix} s & -1 \\ 0 & s \end{pmatrix} \xi_2 = \begin{pmatrix} 0 \\ 1/m \end{pmatrix} F \\ \qquad\qquad y = \begin{pmatrix} 1 & 0 \end{pmatrix} \xi_2 \; , \end{cases} \tag{8.7}$$

where the partial state $\xi_2 = (z, \dot{z})^\top$ is now two-dimensional. △

Similarly to the classical state-space approach, it is also possible, from the polynomial representation, to define the notions of equivalence, observer polynomial form and controller polynomial form to test the properties of controllability and observability of the l.c.s. system (8.2). Before defining these, we introduce some useful notions about polynomial matrices.

8.3 Results on Polynomial Matrices

Notice that the set of polynomial matrices is a noncommutative ring for the addition and multiplication of matrices.

Definition 8.9 *A square polynomial matrix is said to be* regular *if its determinant is a non-identically zero polynomial.*

In the sequel, the matrix $P(s)$ in (8.2) is assumed to be regular.

Definition 8.10 *A square polynomial matrix is said to be* unimodular *if its determinant is equal to a nonzero constant.*

Remark 8.11 The unimodular matrices constitute the invertible elements of the ring of polynomial matrices. ◇

The role of unimodular polynomial matrices is important to obtain particular forms of polynomial matrices and to solve so-called *Bezout identities* as shown below.

8.3.1 Elementary Operations: Hermite and Smith Matrices

It can be shown that every unimodular polynomial matrix can be decomposed into a finite product of elementary matrices corresponding to elementary operations on rows or columns. The rows or columns transformations are classically used in the case of scalar matrices and are basic tools for triangularization of constant matrices, by the Gauss method for example. The same methods apply to polynomial matrices and lead to Hermite and Smith forms. Let us now introduce the elementary unimodular polynomial matrices.

- For each couple (i, j), the matrix P_{ij} is defined by:

$$(P_{ij})_{kl} = \begin{cases} 1 \text{ if } k = l \text{ and } k \neq i \text{ and } k \neq j \\ 1 \text{ if } k = i \text{ and } l = j \\ 1 \text{ if } k = j \text{ and } l = i \\ 0 \text{ else.} \end{cases} \tag{8.8}$$

  Notice that, left-multiplying a polynomial matrix $M(s)$ (respectively right -multiplying) by P_{ij} is equivalent to exchanging the rows i and j (respectively the columns) of $M(s)$.
- For every polynomial $q(s)$, the polynomial matrix $M_i\,[q]\,(s)$ is defined by:

$$\left(M_i\,[q]\right)_{kl}(s) = \begin{cases} 1 & \text{if } k = l \text{ and } k \neq i \\ q(s) & \text{if } k = l = i \\ 0 & \text{else.} \end{cases} \tag{8.9}$$

  In the same way, left-multiplying a polynomial matrix $M(s)$ (respectively right-multiplying) by $M_i\,[q]$ is equivalent to multiplying the ith row of $M(s)$ (respectively column) by $q(s)$.

- For every polynomial $q(s)$, the matrix polynomial $S_{ij}[q]$ is defined by:

$$(S_{ij}[q])_{kl}(s) = \begin{cases} 1 & \text{if } k = l \\ q(s) & \text{if } k = i \text{ and } l = j \\ 0 & \text{else.} \end{cases} \tag{8.10}$$

Left-multiplying a polynomial matrix $M(s)$ (respectively right-multiplying) by $S_{ij}[q]$ is equivalent to adding to the ith row of $M(s)$ (respectively to the jth column) the jth row of $M(s)$ (respectively to the ith column) multiplied by $q(s)$.

We now introduce the so-called *Hermite and Smith forms*.

Definition 8.12 *Let $N(s) = (N_{ij}(s))$ be a polynomial $m \times n$ matrix. The polynomial matrix $N(s)$ is said to be in* upper Hermite form *if:*

1. $N_{ij}(s) = 0$ *if* $i > j$;
2. $\deg(N_{ij}(s)) < \deg(N_{jj}(s))$ *if* $i < j$ *and* $N_{jj}(s) \neq 0$, *else the* $N_{ij}(s)$ *are all equal to zero;*
3. *if* $N_{ii}(s)$ *is nonzero, the coefficient of its highest degree term is equal to 1.*

The polynomial matrix $N(s)$ is said to be in lower Hermite form *if its transpose matrix is in Hermite upper form.*

Definition 8.13 *Let $N(s) = (N_{ij}(s))$ be a polynomial $m \times n$ matrix. The polynomial matrix $N(s)$ is said to be in* Smith form *if:*

1. $N_{ij}(s) = 0$ *for* $i \neq j$;
2. $N_{ii}(s)$ *divides* $N_{i+1,i+1}(s)$.

Definition 8.14 *Let $P_1(s)$ and $P_2(s)$ be two polynomial matrices.*

- *The polynomial matrix $P_2(s)$ is said* row-equivalent *to $P_1(s)$ if $P_2(s) = U_m(s)P_1(s)$, where $U_m(s)$ is a unimodular polynomial matrix.*
- *The polynomial matrix $P_2(s)$ is said* column-equivalent *to $P_1(s)$ if $P_2(s) = P_1(s)U_m(s)$, where $U_m(s)$ is a unimodular polynomial matrix.*

Since all elementary rows or columns operations correspond to multiplications by elementary matrices, we obtain the following proposition (see for example [44, 55, 69]).

Proposition 8.15 *For every polynomial matrix $N(s)$, the following properties are satisfied:*

1. *the polynomial matrix $N(s)$ is row-equivalent to an upper Hermite matrix;*
2. *the polynomial matrix $N(s)$ is column-equivalent to a lower Hermite matrix;*
3. *there exist unimodular polynomial matrices $U_m(s)$ and $V_m(s)$ and a diagonal polynomial matrix $\Gamma(s)$ such that $N(s) = U_m(s)\Gamma(s)V_m(s)$, where $\Gamma(s)$ is said to be the* Smith form *of $N(s)$.*

Example 8.16 Computation of the Smith form of a polynomial matrix.

Let us compute the Smith form of the following polynomial matrix:

$$N(s) = \begin{pmatrix} s-a & 1 \\ 0 & s-a \end{pmatrix}.$$

The procedure is inspired by Proposition 8.15.

The columns are exchanged, which is equivalent to right-multiplying by P_{12} in (8.8), yielding

$$N_1(s) = \begin{pmatrix} 1 & s-a \\ s-a & 0 \end{pmatrix}.$$

The first column of $N_1(s)$ multiplied by $-(s-a)$ is added to the second column of N_1, which is equivalent to right-multiplying N_1 by $S_{12}(-(s-a))$ in (8.10), yielding

$$N_2(s) = \begin{pmatrix} 1 & 0 \\ s-a & -(s-a)^2 \end{pmatrix}.$$

The second column of N_2 is multiplied by -1, which is equivalent to right-multiplying N_2 by $M_2(-1)$ in (8.9), yielding

$$N_3(s) = \begin{pmatrix} 1 & 0 \\ s-a & (s-a)^2 \end{pmatrix}.$$

The first row of N_3 multiplied by $-(s-a)$ is added to its second row, which is equivalent to left-multiplying N_3 by $S_{21}(-(s-a))$, yielding

$$N_4(s) = \begin{pmatrix} 1 & 0 \\ 0 & (s-a)^2 \end{pmatrix}.$$

The Smith form of $N(s)$ is then $N_4(s)$ and we have that $N(s) = U_m(s)N_4(s)V_m(s)$ with:

$$\begin{cases} U_m(s) = \big(S_{21}(-(s-a))\big)^{-1} = \begin{pmatrix} 1 & 0 \\ s-a & 1 \end{pmatrix} \\ V_m(s) = \big(P_{12}S_{12}(-(s-a))M_2(-1)\big)^{-1} = \begin{pmatrix} s-a & 1 \\ -1 & 0 \end{pmatrix}. \end{cases}$$

△

Let us now study the divisibility properties of polynomial matrices.

8.3.2 Division and Bezout Identities

As for the set of polynomials in one variable, the notion of Euclidean division can be introduced in the ring of polynomial matrices. However, this ring being noncommutative, it is necessary to distinguish between left and right divisions. Let us recall, for example, the right divisibility result [34, 35].

Theorem 8.17 *Let $P(s)_{m\times\overline{n}}$ and $Q(s)_{\overline{n}\times\overline{n}}$ be two regular polynomial matrices. Then, there exist two polynomial matrices $D(s)$ and $R(s)$ of dimension $m\times\overline{n}$ such that*

$$P(s) = D(s)Q(s) + R(s) \quad with \quad R(s)Q^{-1}(s)\ strictly\ proper, \tag{8.11}$$

and these matrices are unique.

Definition 8.18 *Let $P(s)$ and $Q(s)$ be two polynomial matrices with the same number of columns (respectively, of rows). The polynomial matrix $D(s)$ is called* right HCD(P, Q) *(highest common divisor) (respectively,* left HCD(P, Q)*) if every right divisor (respectively, left) of P and Q right-divides D (respectively, left-divides).*

*The polynomial matrices $P(s)$ and $Q(s)$ are said to be right-*coprime *(respectively, left) if their right* HCD *(respectively,* left HCD*) is equal to the identity matrix.*

Remark 8.19 A right or left HCD is defined modulo a multiplication by a unimodular polynomial matrix. ◇

The following theorem can then be obtained (see for example [34]).

Theorem 8.20 *Let $P_{\overline{n}\times\overline{n}}$ and $Q_{p\times\overline{n}}$ (respectively, $Q_{\overline{n}\times p}$) be two polynomial matrices. We have that:*

1. $\begin{pmatrix} P \\ Q \end{pmatrix}$ *is row-equivalent to* $\begin{pmatrix} D \\ 0 \end{pmatrix}$, *where D is a right* HCD *of P and Q;*
2. $\begin{pmatrix} P & Q \end{pmatrix}$ *is column-equivalent to* $\begin{pmatrix} D & 0 \end{pmatrix}$, *where D is a left* HCD *of P and Q.*

The following corollary is then straightforward.

Corollary 8.21 *Let P and Q be two left-coprime polynomial matrices, P being regular. There exists a unimodular polynomial matrix U_m such that:*

$$\begin{pmatrix} P & Q \end{pmatrix} U_m = \begin{pmatrix} I & 0 \end{pmatrix}. \tag{8.12}$$

In particular, the following generalized Bezout identities *or* direct Bezout identities *are satisfied: there exist polynomial matrices A, B, C, D, S and T such that*

$$\begin{cases} PA + QB = I \\ PC = QD \\ SA = TB \\ SC + TD = I\ . \end{cases} \tag{8.13}$$

Proof Since P and Q are left-coprime polynomial matrices, from Theorem 8.20 there exists a unimodular matrix U_m satisfying (8.12). Let us partition U_m as follows:

$$U_m = \begin{pmatrix} A & C \\ B & -D \end{pmatrix}.$$

The relation $\begin{pmatrix} P & Q \end{pmatrix} U_m = \begin{pmatrix} I & 0 \end{pmatrix}$ then leads to $PA + QB = I$ and $PC = QD$.

Let us set

$$V_m = U_m^{-1} = \begin{pmatrix} P & Q \\ S & -T \end{pmatrix}.$$

By writing $V_m U_m = I$, we obtain that

$$\begin{pmatrix} P & Q \\ S & -T \end{pmatrix} \begin{pmatrix} A & C \\ B & -D \end{pmatrix} = \begin{pmatrix} I & 0 \\ 0 & I \end{pmatrix}, \tag{8.14}$$

which are nothing but the generalized direct Bezout identities (8.13) (see for example [53]). □

Remark 8.22 Writing $U_m V_m = I$, we obtain the *inverse Bezout identities*. In the same way, if P and Q are two right-coprime polynomial matrices, P being regular, there exists a unimodular polynomial matrix U_m such that $U_m \begin{pmatrix} P \\ Q \end{pmatrix} = \begin{pmatrix} I \\ 0 \end{pmatrix}$. The relations $V_m U_m = I$ and $U_m V_m = I$ define direct and inverse Bezout identities. ◇

These direct and inverse Bezout identities are useful to study the equivalence, controllability, and observability properties. Before we do that, let us introduce the notions of poles and zeros of a transfer matrix using its Smith form.

8.4 Poles and Zeros. Stability

Consider the l.c.s. system described by the polynomial representation (8.2), whose transfer matrix is given by (8.6), that is $H(s) = R(s)P^{-1}(s)Q(s)$.

Definition 8.23 *The* poles *of the l.c.s. system* (8.2) *are the roots of the determinant of* $P(s)$.

This definition has to be related to Definition 5.57.

Proposition 8.24 *If the roots of the determinant of* $P(s)$ *have a strictly negative real part (or are strictly inside the unit disk in the discrete-time case), the l.c.s. system* (8.2) *is* BIBO-*stable.*

The poles of an l.c.s. system determine its stability, whereas the zeros, as it has already been mentioned in § 3.5, play a role with respect to the disturbance rejection problem. Let us define the zeros of the l.c.s. system (8.2) given in polynomial form.

The transfer matrix $H(s)$ in (8.6) of a controllable and observable system can always be written in the following form

$$H(s) = \frac{N(s)}{d(s)} , \tag{8.15}$$

where $d(s)$ is a polynomial in s (and not a polynomial matrix), equal to the smallest common multiple of the denominators of the rational functions elements of $H(s)$ ($d(s)$ can always be chosen such that the coefficient of its highest degree term is equal to 1) and $N(s)$ is a polynomial matrix. Now, from Proposition 8.15, we know that there exist unimodular polynomial matrices $U_m(s)$, $V_m(s)$ and a diagonal matrix $\Gamma(s)$ such that

$$N(s) = U_m(s)\Gamma(s)V_m(s) , \tag{8.16}$$

where $\Gamma(s)$ is the Smith form of $N(s)$. Consequently, we have that

$$U_m^{-1}(s)H(s)V_m^{-1}(s) = \frac{\Gamma(s)}{d(s)} = \text{Diag}(\frac{\gamma_i(s)}{d(s)}) . \tag{8.17}$$

If the rational functions $\gamma_i(s)/d(s)$ are reduced to irreducible rational functions $\varepsilon_i(s)/\psi_i(s)$ for $i = 1, \ldots, r$ (r being the generic rank of $H(s)$), one can then write:

$$H(s) = U_m(s)S(s)V_m(s) \text{ with } S(s) = \begin{pmatrix} \text{Diag}(\frac{\varepsilon_i(s)}{\psi_i(s)}, i = 1, \ldots, r) & 0 \\ 0 & 0 \end{pmatrix} . \tag{8.18}$$

Definition 8.25 *The polynomial matrix $S(s)$ in* (8.18) *is called the* Smith-MacMillan form *of $H(s)$. The* zeros *of $H(s)$, also called* transmission zeros, *are the roots of the $\varepsilon_i(s)$.*

REMARKS

- The following properties are satisfied:
  - $\psi_{i+1}(s)$ divides $\psi_i(s)$, $i = 1, \ldots, r - 1$;
  - $\varepsilon_i(s)$ divides $\varepsilon_{i+1}(s)$, $i = 1, \ldots, r - 1$;
  - $d(s) = \psi_1(s)$.
- As $\varepsilon_i(s)$ divides $\varepsilon_{i+1}(s)$ for $i = 1, \ldots, r - 1$, it can be deduced that the zeros of $H(s)$ are the roots of $\varepsilon_r(s)$. It can also be noticed that the transmission zeros are the complex numbers that make the rank of the matrix $H(s)$ drop.

A dynamical interpretation of the transmission zeros is the following. If s_0 is such a zero, then for an input $u(t) = u_0 \exp(s_0 t)$, $t \geq 0$, there exists an initial state x_0 such that the output $y(t)$ is zero for $t \geq 0$ (see [44]).

Example 8.26 Consider the scalar l.c.s. system with transfer function

$$H(s) = \frac{s - s_0}{s + a}, \quad a > 0 .$$

The system has the pole $-a < 0$ and, therefore, it is BIBO-stable by Proposition 8.24. The system is not strictly causal since the numerator and the denominator have the same degree: there is a direct link between the input and the output. It can be written for example in the following state-space form as

$$\begin{cases} \dot{x} = -ax + u \\ y = -(s_0 + a)x + u , \end{cases}$$

with the corresponding time solution

$$x(t) = e^{-at}x_0 + \int_0^t e^{-a(t-\tau)}u(\tau)\,\mathrm{d}\tau .$$

Replacing $u(\tau)$ by $u_0 e^{s_0\tau}$ in the previous equation, we obtain:

$$y(t) = e^{-at}(u_0 - (s_0 + a)x_0) .$$

It can be observed that, for the initial state $x_0 = \dfrac{u_0}{s+a}$, the output $y(t)$ is identically zero for $t \geq 0$. △

8.5 Equivalence Between Linear Differential Systems

We first enounce without demonstration a technical result which gives a necessary and sufficient condition for two systems of linear differential equations to have the "same solutions", that is the same sets of solutions or isomorphic trajectories (we refer the reader to [36, part IV, Chap. S1]).

Proposition 8.27 *Let $A(s)$ and $\widetilde{A}(s)$ be two regular polynomial $n \times n$ matrices and $B(s)$ and $\widetilde{B}(s)$ be two polynomial $n \times m$ matrices. The two systems of controlled linear differential equations*

$$A(s)x = B(s)u , \quad \widetilde{A}(s)\tilde{x} = \widetilde{B}(s)u ,$$

have the same solutions if, and only if, the matrices $(A(s), B(s))$ and $(\widetilde{A}(s), \widetilde{B}(s))$ are row-equivalent, namely if and only if there exists a unimodular polynomial $n \times n$ matrix U_m such that:

$$\begin{cases} \widetilde{A}(s) = U_m(s)A(s) \\ \widetilde{B}(s) = U_m(s)B(s) . \end{cases} \tag{8.19}$$

Let us now introduce the notion of equivalence of two systems in polynomial representation.

Definition 8.28 *Let* (Σ_1) *and* (Σ_2) *be two input-output l.c.s. systems in polynomial form with the same input and output vectors* u *and* y*:*

$$(\Sigma_1) \qquad \begin{cases} P_1(s)\xi_1 = Q_1(s)u \\ \qquad\; y = R_1(s)\xi_1 \end{cases} \tag{8.20}$$

and

$$(\Sigma_2) \qquad \begin{cases} P_2(s)\xi_2 = Q_2(s)u \\ \qquad\; y = R_2(s)\xi_2 \; . \end{cases} \tag{8.21}$$

The partial state ξ_1 *is of dimension* n_1 *and* ξ_2 *of dimension* n_2*. Then* (Σ_1) *and* (Σ_2) *are said to be* equivalent *if there exist polynomial matrices* $M_1(s)_{n_1 \times n_2}$, $M_2(s)_{n_2 \times n_1}$, $N_1(s)_{n_1 \times m}$ *and* $N_2(s)_{n_2 \times m}$ *such that the two following systems* (S_1) *and* (S_2) *have the same solutions:*

$$(S_1) \qquad \begin{cases} P_1(s)\xi_1 = Q_1(s)u \\ \qquad\;\; \xi_2 = M_2(s)\xi_1 + N_2(s)u \\ \qquad\;\; y = R_1(s)\xi_1 \end{cases} \tag{8.22}$$

$$(S_2) \qquad \begin{cases} \qquad\;\; \xi_1 = M_1(s)\xi_2 + N_1(s)u \\ P_2(s)\xi_2 = Q_2(s)u \\ \qquad\;\; y = R_2(s)\xi_2 \; . \end{cases} \tag{8.23}$$

Remark 8.29 In the classical state-space representation, the equivalence property is characterized by the transformation of state vectors by simple changes of basis (see Definition 5.18), whereas here the transformation from ξ_1 to ξ_2 (respectively from ξ_2 to ξ_1) uses polynomial matrices (and no longer constant matrices) as well as the control input u. ◇

Denoting $A(s)$ and $B(s)$ the polynomial matrices associated with Eq. (8.22), and $\widetilde{A}(s)$ and $\widetilde{B}(s)$ those associated with Eq. (8.23), we have that:

$$\begin{cases} A(s) = \begin{pmatrix} P_1 & 0 & 0 \\ -M_2 & I & 0 \\ -R_1 & 0 & I \end{pmatrix}, \quad B(s) = \begin{pmatrix} Q_1 \\ N_2 \\ 0 \end{pmatrix} \\ \widetilde{A}(s) = \begin{pmatrix} I & -M_1 & 0 \\ 0 & P_2 & 0 \\ 0 & -R_2 & I \end{pmatrix}, \quad \widetilde{B}(s) = \begin{pmatrix} N_1 \\ Q_2 \\ 0 \end{pmatrix} . \end{cases} \tag{8.24}$$

By Proposition 8.27, if the l.c.s. systems (S_1) and (S_2) have the same solutions, there exists a unimodular polynomial matrix $U_m(s)$ satisfying (8.19). From the particular structures of $A(s)$ and $\widetilde{A}(s)$, it can be deduced that $U_m(s)$ and $U_m^{-1}(s)$ have the following form:

$$U_m(s) = \begin{pmatrix} S & -M_1 & 0 \\ T & P_2 & 0 \\ V & -R_2 & I \end{pmatrix}, \quad U_m^{-1}(s) = \begin{pmatrix} P_1 & C & 0 \\ -M_2 & D & 0 \\ -R_1 & E & I \end{pmatrix}. \tag{8.25}$$

Example 8.30 Let us consider again the example of the material point discussed in § 8.3. Systems (8.4) and (8.7) are equivalent since we have that

$$\xi_2 = \begin{pmatrix} 1 \\ s \end{pmatrix} \xi_1 , \quad \xi_1 = \begin{pmatrix} 1 & 0 \end{pmatrix} \xi_2 ,$$

and, from (8.25), we can write

$$U_m(s) = \begin{pmatrix} 0 & -1 & 0 & 0 \\ 0 & s & -1 & 0 \\ 1 & 0 & s & 0 \\ 0 & -1 & 0 & 1 \end{pmatrix}.$$

It can be easily checked that polynomial matrices $A(s)$, $\widetilde{A}(s)$, $B(s)$, $\widetilde{B}(s)$ given by (8.24) can be written as

$$\begin{cases} A(s) = \begin{pmatrix} s^2 & 0 & 0 & 0 \\ -1 & 1 & 0 & 0 \\ -s & 0 & 1 & 0 \\ -1 & 0 & 0 & 1 \end{pmatrix}, & B(s) = \begin{pmatrix} 1/m \\ 0 \\ 0 \\ 0 \end{pmatrix} \\ \widetilde{A}(s) = \begin{pmatrix} 1 & -1 & 0 & 0 \\ 0 & s & -1 & 0 \\ 0 & 0 & s & 0 \\ 0 & -1 & 0 & 1 \end{pmatrix}, & \widetilde{B}(s) = \begin{pmatrix} 0 \\ 0 \\ 1/m \\ 0 \end{pmatrix}, \end{cases}$$

and readily satisfy equalities (8.19). △

8.6 Observability and Controllability

We can now introduce the controllability and observability notions in the context of polynomial representation. The links with the state-space representation are also established.

8.6.1 Controllability

Definition 8.31 *The l.c.s. system (Σ), given by* (8.2), *is said to be* controllable *if the polynomial matrices $P(s)$ and $Q(s)$ are left-coprime.*

Let us motivate this definition. By Definition 8.18, if P and Q are not left-coprime, there exists a polynomial matrix $D(s)$ with $\deg\big(\det(D)\big) \geq 1$ such that:

$$P = DP_1 \quad \text{and} \quad Q = DQ_1 \ .$$

The equation $P\xi = Qu$ then becomes $D(P_1\xi - Q_1 u) = 0$. If we denote $\zeta = P_1\xi - Q_1 u$, then ζ satisfies the following differential equation

$$D(s)\zeta = 0 \ ,$$

and the dynamics of ζ only depends on the initial conditions of the system, and is therefore independent of the control u: the "kernel" of the operator D represents a non controllable subspace.

Let us now introduce the notion of *polynomial controller form*, counterpart of the canonical controllable form in the classical state-space context (see § 5.4.2).

Definition 8.32 *An l.c.s. system is said to be in the* polynomial controller form *if it is in the form*

$$\begin{cases} P_c(s)\xi = u \\ \quad\quad y = R_c(s)\xi \ , \end{cases} \tag{8.26}$$

or, in other words, if $Q(s) = I$ in (8.2).

This controller form is well adapted to the computation of the closed-loop system's behavior. The partial state ξ, which will be used as the argument of a feedback to make possible a pole placement (as explained below), has the same dimension as the control u, the matrix $Q_c(s)$ being here equal to the identity matrix.

Let us now establish the following theorem justifying the notion of polynomial controller form.

Theorem 8.33 *Let (Σ) be an l.c.s. system in the polynomial form* (8.2). *The following assertions are equivalent:*

(i) (Σ) is controllable;
(ii) (Σ) is equivalent to a polynomial controller form.

Proof **ii)** $\rightarrow$ **i)** (Σ) is assumed to be equivalent to the following polynomial controller form:

$$\begin{cases} P_c(s)\zeta = u \\ \quad\quad y = R_c(s)\zeta \ . \end{cases} \tag{8.27}$$

From Definition 8.28, the existence of polynomial matrices M_1, N_1, M_2 and N_2 of suitable dimensions and of a unimodular polynomial matrix U_m of the form (8.25) can be deduced, namely

$$U_m = \begin{pmatrix} S & -M_1 & 0 \\ T & P_c & 0 \\ V & -R_c & I \end{pmatrix} \tag{8.28}$$

such that $U_m \begin{pmatrix} A & B \end{pmatrix} = \begin{pmatrix} \widetilde{A} & \widetilde{B} \end{pmatrix}$, which gives:

$$\begin{pmatrix} S & -M_1 & 0 \\ T & P_c & 0 \\ V & -R_c & I \end{pmatrix} \begin{pmatrix} P & 0 & 0 & Q \\ -M_2 & I & 0 & N_2 \\ -R & 0 & I & 0 \end{pmatrix} = \begin{pmatrix} I & -M_1 & 0 & N_1 \\ 0 & P_c & 0 & I \\ 0 & -R_c & I & 0 \end{pmatrix}.$$

By the associativity property, we have that

$$\begin{pmatrix} I & -N_1 \\ 0 & I \end{pmatrix} \begin{pmatrix} S & M_1 \\ T & -P_c \end{pmatrix} \begin{pmatrix} P & Q \\ M_2 & -N_2 \end{pmatrix} = \begin{pmatrix} I & -N_1 \\ 0 & I \end{pmatrix} \begin{pmatrix} I & N_1 \\ 0 & I \end{pmatrix} = \begin{pmatrix} I & 0 \\ 0 & I \end{pmatrix}.$$

Computing first the product $\begin{pmatrix} I & -N_1 \\ 0 & I \end{pmatrix} \begin{pmatrix} S & M_1 \\ T & -P_c \end{pmatrix}$ leads to:

$$W_m \begin{pmatrix} P & Q \\ M_2 & -N_2 \end{pmatrix} = \begin{pmatrix} I & 0 \\ 0 & I \end{pmatrix} \quad \text{with } W_m = \begin{pmatrix} S - N_1 T & M_1 + N_1 P_c \\ T & -P_c \end{pmatrix}. \tag{8.29}$$

But $\begin{pmatrix} I & -N_1 \\ 0 & I \end{pmatrix}$ is unimodular and, using the expression (8.28) of U_m, the same holds true for $\begin{pmatrix} S & M_1 \\ T & -P_c \end{pmatrix}$. Consequently, W_m is unimodular as the product of two unimodular polynomial matrices and, from (8.29), it can be deduced that $W_m^{-1} = \begin{pmatrix} P & Q \\ M_2 & -N_2 \end{pmatrix}$, viz.

$$\begin{pmatrix} P & Q \\ M_2 & -N_2 \end{pmatrix} W_m = \begin{pmatrix} I & 0 \\ 0 & I \end{pmatrix}. \tag{8.30}$$

Equality (8.29) (respectively, (8.30)) is the direct Bezout identity (respectively, inverse) of Corollary 8.21, associated with the matrices P and Q which are therefore left-coprime, which in turn implies the controllability property of system (Σ).

i) $\rightarrow$ **ii)** (Σ) is assumed to be controllable. Therefore, by Definition 8.31 and Corollary 8.21, there exist Bezout relations between P and Q of the form

$$\begin{pmatrix} P & Q \\ S & -T \end{pmatrix} \begin{pmatrix} A & C \\ B & -D \end{pmatrix} = \begin{pmatrix} A & C \\ B & -D \end{pmatrix} \begin{pmatrix} P & Q \\ S & -T \end{pmatrix} = \begin{pmatrix} I & 0 \\ 0 & I \end{pmatrix}$$

which implies:

$$\begin{pmatrix} A & C & 0 \\ B & -D & 0 \\ RA & RC & I \end{pmatrix} \begin{pmatrix} P & 0 & 0 & Q \\ S & I & 0 & -T \\ -R & 0 & I & 0 \end{pmatrix} = \begin{pmatrix} I & C & 0 & 0 \\ 0 & D & 0 & I \\ 0 & RC & I & 0 \end{pmatrix}.$$

Then the following equalities can be deduced:

$$\begin{cases} P\xi = Qu \\ y = R\xi \\ \zeta = -S\xi - Tu \end{cases} \qquad \begin{cases} -D\zeta = u \\ y = -RC\zeta \\ \xi = -C\zeta \ . \end{cases}$$

This means that (Σ) is equivalent to the following system in polynomial controller form

$$\begin{cases} -D\zeta = u \\ y = -RC\zeta \ , \end{cases}$$

which concludes the proof. □

Let us now prove that the controllability notions coincide for systems given in a classical state-space form or in a polynomial form, by first stating the following technical lemma.

Lemma 8.34 *Let A be a real $n \times n$ matrix and B a real $n \times m$ matrix. The two following conditions are equivalent:*

(i) $\operatorname{rank}\left(B, AB, \ldots, A^{n-1}B\right) = n$*;*
(ii) the polynomial matrices $(sI - A)$ and B are left-coprime.

Proof **ii)** $\to$ **i)** The polynomial matrices $(sI - A)$ and B being left-coprime, there exists, by Corollary 8.21, a left Bezout relation of the form

$$(sI - A)P(s) + BQ(s) = I,$$

where P and Q are polynomial matrices. If these matrices are expressed as $P = \sum_{i=0}^{r} P_i s^i$ and $Q = \sum_{i=0}^{r} Q_i s^i$, we can write:

$$(sI - A)P + BQ = P_r s^{r+1} + \sum_{i=1}^{r} (P_{i-1} - AP_i + BQ_i)\, s^i + (BQ_0 - AP_0) \ .$$

Then the following equalities can be deduced

$$\begin{cases} P_r &= 0 \\ P_{r-1} &= -BQ_r \\ P_{r-2} &= -BQ_{r-1} - ABQ_r \\ \cdots &= \cdots \\ P_0 &= -BQ_1 - ABQ_2 - \cdots - A^{r-1}BQ_r \ , \end{cases}$$

as well as

$$I = BQ_0 - AP_0 = BQ_0 + ABQ_1 + \cdots + A^r BQ_r$$

$$= \left(B,\, AB,\, \cdots,\, A^r B\right) \begin{pmatrix} Q_0 \\ Q_1 \\ \vdots \\ Q_r \end{pmatrix}. \tag{8.31}$$

Therefore, the matrix $\left(B,\, AB,\, \ldots,\, A^{r-1}B\right)$ is right-invertible. In particular, it has rank n and Cayley-Hamilton theorem 4.10 makes it possible to conclude that $\operatorname{rank}\left(B,\, AB,\, \ldots,\, A^{n-1}B\right) = n$.

i) → **ii)** Condition **i)** implies the existence of constant matrices $Q_0, \ldots, Q_{n-1}$ such that (8.31) is satisfied with $r = n-1$. Then the polynomial matrices $P = \sum_{i=0}^{n-2} P_i s^i$ and $Q = \sum_{i=0}^{n-1} Q_i s^i$ make it possible to establish a left Bezout relation between $(sI - A)$ and B, which gives **i)** and concludes the proof of the lemma. □

The following theorem is a straightforward consequence of Theorem 5.12.

Theorem 8.35 *The l.c.s. system in state-space representation* $\dot{x} = Ax + Bu$ *is controllable if and only if the polynomial matrices* $(sI - A)$ *and* B *are left-coprime.*

8.6.2 Observability

Let us now turn to the dual property of observability.

Definition 8.36 *The l.c.s. system* (Σ)*, given by (8.2), is said to be* observable *if the polynomial matrices* $P(s)$ *and* $R(s)$ *are right-coprime.*

Let us motivate this definition. Saying that $P(s)$ and $R(s)$ are right-coprime means that, using the inverse Bezout identities (see Remark 8.22), we have that $U_m \begin{pmatrix} P \\ R \end{pmatrix} = \begin{pmatrix} I \\ 0 \end{pmatrix}$ with a unimodular polynomial matrix of the form $U_m = \begin{pmatrix} A & B \\ C & -D \end{pmatrix}$. Then, the following equalities are satisfied:

$$AP + BR = I \text{ and } CP = DR\,.$$

Right-multiplying the first equality by ξ, we obtain from (8.2),

$$AQu + By = \xi\,,$$

and y and u make it possible to reconstitute the partial state ξ, which is nothing but the mathematical expression of the observability property. Then, if the output is equal to the partial state, the observability property is straightforward and this justifies the following definition.

Definition 8.37 *An l.c.s. system is said to be in* polynomial observer form *if it is in the form:*

$$\begin{cases} P_o(s)\xi = Q_o(s)u \\ y = \xi \,. \end{cases} \tag{8.32}$$

Notice that it means that the output is equal to the partial state.

As for the controllability property, we have the following theorem.

Theorem 8.38 *Let (Σ) be a system in polynomial form (8.2). The following propositions are equivalent:*

(i) (Σ) is observable;
(ii) (Σ) is equivalent to the polynomial observer form.

Proof **ii)** $\rightarrow$ **i)** We suppose that (Σ) is equivalent to the following polynomial observer form

$$P_o y = Q_o u \,. \tag{8.33}$$

From Definition 8.28, there exist polynomial matrices M, N, S and T and a unimodular polynomial matrix U_m of the form

$$U_m = \begin{pmatrix} U_{m,1} & U_{m,2} & 0 \\ U_{m,3} & U_{m,4} & 0 \\ U_{m,5} & U_{m,6} & I \end{pmatrix}$$

such that:

$$U_m \begin{pmatrix} P & 0 & 0 & Q \\ -S & I & 0 & T \\ -R & 0 & I & 0 \end{pmatrix} = \begin{pmatrix} I & -M & 0 & N \\ 0 & P_o & 0 & Q_o \\ 0 & -I & I & 0 \end{pmatrix}.$$

In particular, we deduce that

$$\begin{cases} U_{m,6} = -I \\ U_{m,1}P - U_{m,2}S = I \\ U_{m,5}P - U_{m,6}S - R = 0 \,. \end{cases}$$

These equalities imply that

$$(U_{m,1} + U_{m,2}U_{m,5})P - U_{m,2}R = I \,. \tag{8.34}$$

From (8.34), it can be deduced that P are R are right-coprime and, therefore, that (Σ) is observable.

i) $\rightarrow$ **ii)** Let us now suppose that P and R are right-coprime. Therefore, by Corollary 8.21, we can write the Bezout relations

$$\begin{pmatrix} A & B \\ C & -D \end{pmatrix}\begin{pmatrix} P & S \\ R & -T \end{pmatrix} = \begin{pmatrix} P & S \\ R & -T \end{pmatrix}\begin{pmatrix} A & B \\ C & -D \end{pmatrix} = \begin{pmatrix} I & 0 \\ 0 & I \end{pmatrix},$$

which imply that

$$AP + BR = I \; .$$

Multiplying by ξ and using (8.2) this gives $AQu + By = \xi$. Therefore, the changes of partial states

$$\begin{cases} y = R\xi \\ \xi = AQu + By \end{cases}$$

induce an equivalence (in the polynomial sense) between (Σ) and (8.33), which concludes the proof. □

Similarly to the controllability case, it can be shown that the observability notions coincide for polynomial systems and systems in classical state-space representation. The following lemma is satisfied.

Lemma 8.39 *Let A be a real $n \times n$ matrix and C a real $p \times n$ matrix. The two following conditions are equivalent:*

(i) $\operatorname{rank}\begin{pmatrix} C \\ CA \\ \vdots \\ CA^{n-1} \end{pmatrix} = n;$

(ii) the polynomial matrices $(sI - A)$ and C are right-coprime.

Proof The proof, quite similar to the one of Lemma 8.34, is left as an exercise for the reader. □

The following theorem can be deduced.

Theorem 8.40 *The l.c.s. system in state-space representation $\dot{x} = Ax + Bu$ with the output $y = Cx$ is observable if and only if the polynomial matrices $(sI - A)$ and C are right-coprime.*

REMARKS

- In the scalar case, the transfer function of system (8.2) can be written as:

$$h(s) = \frac{r(s)q(s)}{p(s)} \; .$$

The observability property means that the polynomials $p(s)$ and $r(s)$ are coprime and the controllability means that $p(s)$ are $q(s)$ coprime. If the system is not observable (respectively, not controllable), the unobservable modes (respectively, uncontrollable) do not appear in the expression of the transfer function, as we have seen in Propositions 5.60 and 5.61. A pole/zero simplification necessarily has occurred, which means that $p(s)$ and $r(s)$ (respectively $p(s)$ are $q(s)$) are not coprime.

- If the system is not controllable, the zeros of the Smith form of $(P(s), Q(s))$ correspond to the uncontrollable modes, and are called *input decoupling zeros* (see for example [44, 55]).
  If the system is not observable, the zeros of the Smith form of $\begin{pmatrix} P(s) \\ R(s) \end{pmatrix}$ correspond to the unobservable modes and are called *output decoupling zeros* (see [44, 55]).
- In the sequel, we consider controllable and observable systems. Consequently, when we refer to zeros, we mean transmission zeros , that is, zeros of the transfer matrix (see Definition 8.25). Finally, this assumption implies that the system's poles, equal to the roots of the determinant of $P(s)$, are also given by the roots of the $\psi_i(s)$ of the Smith-Mac-Millan form (8.18).

8.7 From the State-Space Representation to the Polynomial Controller and Observable Forms

We now establish the links between the state-space and the polynomial representations of l.c.s. systems. More precisely, we give a systematic way to compute the polynomial controller and observer forms from the classical state-space representation.

We consider a controllable and observable linear dynamical system (Σ) in state-space form, with an additive dynamics disturbance w of dimension r:

$$(\Sigma) \quad \begin{cases} \dot{x} = Fx + Gu + G_d w \\ y = Cx \ . \end{cases} \tag{8.35}$$

8.7.1 *From the State-Space Representation to the Polynomial Observer Form*

We first state the following proposition giving the polynomial observer form associated with (8.35).

Proposition 8.41 *The polynomial observer form associated with (8.35) is given by*

$$A(s)y = B_c(s)u + B_d(s)w \quad with \quad \begin{cases} B_c(s) = B(s)G \\ B_d(s) = B(s)G_d \end{cases} \tag{8.36}$$

where $A(s)$ and $B(s)$ satisfy

$$A(s)C = B(s)(sI - F) \ .$$

Proof The system (Σ) being observable, the polynomial matrices $(sI - F)$ and C are right-coprime. Let us define

$$\widetilde{F}(s) = sI - F \ .$$

By Corollary 8.21 and Remark 8.22, we know that there exist left-coprime polynomial matrices $A(s)$ and $B(s)$ and $X(s)$, $Y(s)$, $X_1(s)$, $Y_1(s)$ satisfying the direct and inverse matrix Bezout identities, namely:

$$\begin{cases} \begin{pmatrix} A(s) & -B(s) \\ Y_1(s) & X_1(s) \end{pmatrix} \begin{pmatrix} X(s) & C \\ -Y(s) & \widetilde{F}(s) \end{pmatrix} = \begin{pmatrix} I & 0 \\ 0 & I \end{pmatrix} \\ \begin{pmatrix} X(s) & C \\ -Y(s) & \widetilde{F}(s) \end{pmatrix} \begin{pmatrix} A(s) & -B(s) \\ Y_1(s) & X_1(s) \end{pmatrix} = \begin{pmatrix} I & 0 \\ 0 & I \end{pmatrix} . \end{cases} \tag{8.37}$$

Equalities (8.37) imply in particular that

$$Y_1(s)C + X_1(s)\widetilde{F}(s) = I,$$

and, right-multiplying by x this matrix equation, we obtain from (8.35):

$$Y_1(s)y + X_1(s)(Gu + G_d w) = x \ .$$

This equality is nothing but the mathematical expression of the observability property, that is, the fact that the state x can be obtained from the knowledge of the output y and the inputs u and w. From (8.37), we have that:

$$A(s)C = B(s)\widetilde{F}(s) \ .$$

On the other hand, since $y = Cx$, we deduce that

$$A(s)y = B(s)\widetilde{F}(s)x = B(s)Gu + B(s)G_d w \ ,$$

which constitutes the expected polynomial observer form (8.36). □

8.7.2 From the Polynomial Observer form to the Polynomial Controller Form

The following proposition gives the polynomial controller form associated with (8.35).

Proposition 8.42 *The polynomial controller form associated with* (8.35) *is given by*

$$\begin{cases} A_1(s)\xi = u + T_1(s)w \\ \quad\quad y = B_1(s)\xi + T_2(s)w \ , \end{cases} \tag{8.38}$$

$$\textit{with } T_1(s) = F_0(s)B_d(s) \textit{ and } T_2(s) = E_0(s)B_d(s) \ ,$$

where $A_1(s)$*,* $B_1(s)$ *and* $E_0(s)$*,* $F_0(s)$ *satisfy*

$$\begin{cases} A(s)B_1(s) = B_c(s)A_1(s) \\ A(s)E_0(s) + B_c(s)F_0(s) = I \ , \end{cases}$$

the polynomial matrices $A(s)$ *and* $B_c(s)$ *being associated with the polynomial observer form* (8.36) *of* (8.35).

Proof We know that (8.36) is equivalent to the l.c.s. system (8.35) which is controllable by assumption. Consequently, the matrices $A(s)$ and $B_c(s)$ are left-coprime. By Corollary 8.21, we can write again the associated matrix Bezout identities and we know that there exist right-coprime polynomial matrices $A_1(s)$ and $B_1(s)$ and $E_0(s)$, $F_0(s)$, $E_1(s)$, $F_1(s)$ satisfying:

$$\begin{cases} \begin{pmatrix} A(s) & -B_c(s) \\ F_1(s) & E_1(s) \end{pmatrix} \begin{pmatrix} E_0(s) & B_1(s) \\ -F_0(s) & A_1(s) \end{pmatrix} = \begin{pmatrix} I & 0 \\ 0 & I \end{pmatrix} \\ \begin{pmatrix} E_0(s) & B_1(s) \\ -F_0(s) & A_1(s) \end{pmatrix} \begin{pmatrix} A(s) & -B_c(s) \\ F_1(s) & E_1(s) \end{pmatrix} = \begin{pmatrix} I & 0 \\ 0 & I \end{pmatrix} . \end{cases} \tag{8.39}$$

Let ξ be the vector defined by (the s variable is omitted in the sequel):

$$\xi = F_1(y - E_0 B_d w) + E_1(u + F_0 B_d w) \ . \tag{8.40}$$

Then, we have that $A_1\xi = A_1 F_1(y - E_0 B_d w) + A_1 E_1(u + F_0 B_d w)$ and, from (8.39), that

$$A_1\xi = F_0 A(y - E_0 B_d w) + (I - F_0 B_c)(u + F_0 B_d w) \ . \tag{8.41}$$

From the polynomial observer form (8.36), we have that

$$F_0 A(y - E_0 B_d w) = F_0 B_c u + F_0(I - A E_0) B_d w \ ,$$

and also, from (8.39),

$$I - A E_0 = B_c F_0 \ . \tag{8.42}$$

We deduce that

$$F_0 A(y - E_0 B_d w) = F_0 B_c(u + F_0 B_d w) \ ,$$

which implies, replacing $F_0 A(y - E_0 B_d w)$ in (8.41), that:

$$A_1\xi = u + F_0 B_d w \ . \tag{8.43}$$

In the same way, from the expression (8.40) of ξ, we obtain

$$B_1\xi = B_1 F_1(y - E_0 B_d w) + B_1 E_1(u + F_0 B_d w) \ .$$

Now, using again the Bezout identities (8.39), we obtain

$$E_0 A = I - B_1 F_1 \ \text{ and } \ E_0 B_c = B_1 E_1 \ ,$$

which gives:

$$B_1\xi = (I - E_0 A)(y - E_0 B_d w) + E_0 B_c(u + F_0 B_d w) \ . \tag{8.44}$$

On the other hand, we can write from (8.36)

$$E_0 A(y - E_0 B_d w) = E_0 B_c u + E_0(I - A E_0) B_d w \ ,$$

which gives, using equality (8.42),

$$E_0 A(y - E_0 B_d w) = E_0 B_c(u + F_0 B_d w) \ .$$

Replacing $E_0 A(y - E_0 B_d w)$ by its expression in (8.44), we obtain

$$B_1\xi = y - E_0 B_d w \ , \tag{8.45}$$

and considering (8.43) and (8.45), we finally obtain the expected polynomial controller form (8.38). This concludes the proof. □

8.8 Closed-Loop Transfer Functions from the Input and the Disturbances to the Outputs

To be more general, we consider the polynomial controller form (8.38) and we add a vector d_m of measurements disturbances. This form is well adapted to the computation of the closed-loop transfers, since we show now that a feedback law can be expressed directly from the partial state.

Consider the system given by:

$$\begin{cases} A_1(s)\xi = u + F_0 B_d w \\ \quad\ \ y = B_1(s)\xi + E_0 B_d w + d_m \ . \end{cases} \tag{8.46}$$

Let us now define what is meant by "controller" in the polynomial context.

Definition 8.43 *We call* causal linear dynamic rational controller *with input y and output u, a system of the form*

$$P(s)u + Q(s)y = r \,, \tag{8.47}$$

where $r(t)$ is a reference signal, and $P(s)$ and $Q(s)$ are polynomial matrices of suitable dimensions such that $P^{-1}(s)Q(s)$ is proper.

Let us compute the characteristic polynomial of the closed-loop system (8.46)–(8.47). The expressions of y and u given by (8.46) are replaced in (8.47), so that the feedback law can be expressed as follows, using the partial state ξ of the polynomial controller form (8.38) associated with (8.35):

$$r = T\xi + (QE_0B_d - PF_0B_d)w + Qd_m \text{ with } T = PA_1 + QB_1 \,. \tag{8.48}$$

Remark 8.44 The matrix $T(s)$ is invertible since its determinant has for roots the poles of the closed-loop system which are chosen asymptotically stable. ◇

Then, from (8.48), ξ can be written as

$$\xi = T^{-1}(r + (PF_0 - QE_0)B_d w - Qd_m) \,, \tag{8.49}$$

which leads to the following proposition giving the expression of the different transfers between r, w, d_m and y.

Proposition 8.45 *The closed-loop expression of y in function of the reference r, the dynamics disturbances w and the measurements disturbances d_m is of the form:*

$$y = T_{ry}r + T_{wy}w + T_{d_m y}d_m \,,$$

$$\text{with} \begin{cases} T_{ry} &= B_1T^{-1} \\ T_{wy} &= (B_1T^{-1}(PF_0 - QE_0) + E_0)B_d \\ T_{d_m y} &= I - B_1T^{-1}Q \,. \end{cases} \tag{8.50}$$

Remark 8.46 It can be noticed that the zeros of the open-loop system (8.46), that is, the zeros of the transfer from u to y given by $B_1(s)$, are also present after feedback in the transfer T_{ry} [51, 52, 54]. The zeros of the transfer T_{ry} are usually called *tracking zeros*. ◇

Let us now parameterize the controller to make the degrees of freedom appear in an affine way. They make it possible to act on the zeros of the transfer functions T_{wy} and $T_{d_m y}$, which are called *regulation zeros*.

8.9 Affine Parameterization of the Controller and Zeros Placement with Fixed Poles

This kind of parameterization, initially used by Youla et al. [71], is useful to obtain stabilizing controllers optimizing some criteria. We use the same type of parameterization of the controller for a slightly different objective. In fact, the closed-loop poles having been fixed in § 8.8, we have made good use of the degrees of freedom, appearing affinely in the controller's parameterization (8.48), to place some zeros of suitable transfer functions.

Using the fact that the polynomial matrices $A_1(s)$ and $B_1(s)$ in (8.46) are right-coprime, we know that, by Corollary 8.21, there exist polynomial matrices $P_0(s)$ and $Q_0(s)$ such that:

$$P_0 A_1 + Q_0 B_1 = I \; . \tag{8.51}$$

On the other hand, since $AB_1 = B_c A_1$ where A and B_c are associated with the polynomial observer form (8.36), the general solutions P and Q satisfy $T = PA_1 + QB_1$, where $T(s)$ is the matrix associated with the closed-loop dynamics, and are given by

$$\begin{cases} P = TP_0 + KB_c \\ Q = TQ_0 - KA \; , \end{cases} \tag{8.52}$$

where $K(s)$ here denotes an arbitrary polynomial matrix. Using this parameterization, the different transfers of (8.50) can be rewritten as:

$$\begin{cases} T_{ry} &= B_1 T^{-1} \\ T_{wy} &= \Big(B_1\big(P_0F_0 - Q_0E_0 + T^{-1}K(B_cF_0 + AE_0)\big) + E_0\Big)B_d \\ T_{d_m y} &= I - B_1Q_0 + B_1T^{-1}KA \; . \end{cases} \tag{8.53}$$

The following proposition is straightforward.

Proposition 8.47 *Thanks to the degrees of freedom of the matrix K affinely appearing in the polynomial matrices $T_{wy}(s)$ and $T_{d_m y}(s)$ in (8.53), it is possible to act on the regulation zeros appearing after feedback, without modifying the poles characterized by the determinant of $T(s)$.*

To conclude, it has been shown that the state-space representation and the polynomial representation are conceptually equivalent. However, the degrees of freedom potentially existing in the control, in the multivariable case, are not highlighted in the classical state-space stabilization approach through the observer-regulator synthesis. The idea is to use these degrees of freedom from the polynomial controller form, to place some regulation zeros to asymptotically reject some disturbances, with a controller of minimal dimension. This method is applied in the next section devoted to the stabilization of the inverted pendulum.

8.10 The Inverted Pendulum Example

The inverted pendulum on a moving cart with mass has been introduced in § 2.3.1, and described by the controlled nonlinear dynamical system (2.9). The tangent linear dynamical system of (2.9) at the unstable equilibrium is given in (4.38) and in (5.4):

$$\begin{cases} \dot{x} = Fx + Gu \\ y = Cx \ , \end{cases} \tag{8.54}$$

with

$$F = \begin{pmatrix} 0 & 0 & 1 & 0 \\ 0 & 0 & 0 & 1 \\ 0 & a & 0 & 0 \\ 0 & b & 0 & 0 \end{pmatrix}, \quad G = \begin{pmatrix} 0 \\ 0 \\ r_1 \\ r_2 \end{pmatrix}, \quad C = \begin{pmatrix} 1 & 0 & 0 & 0 \\ 0 & 1 & 0 & 0 \end{pmatrix}. \tag{8.55}$$

The coefficients of the matrices F and G are constant and depend on the physical parameters of the system. System (8.54)–(8.55) is controllable and observable and therefore the $n = 4$ poles of $F - GK$ and the $n = 4$ poles of $F - LC$ can be placed separately (see Chaps. 5 and 7). Consequently, the number n_d of degrees of freedom of the controller is given by (8.1), viz.:

$$n_d = n(m + p - 2) = 4 \ .$$

These degrees of freedom are used to place some regulation zeros.

In fact, many disturbances affect the system: measurements disturbances, unknown slope of the bench, motors's dissymmetries, dry friction along the bench, etc. So, to be more complete, a n-vector w of dynamics disturbances and a p-vector d_m of measurements disturbances have been added:

$$w = (0\,0\,w_1\,w_2)^\top, \quad d_m = (bz\,b\theta)^\top. \tag{8.56}$$

8.10.1 Computation of the Polynomial Observer and Controller Forms

Applying the results of § 8.7, the polynomial observer form can be obtained directly from (8.54)–(8.55):

$$A(s) = \begin{pmatrix} s^2 & -a \\ 0 & s^2 - b \end{pmatrix}, \quad B_c(s) = \begin{pmatrix} r_1 \\ r_2 \end{pmatrix}, \quad B_d(s) = I \ . \tag{8.57}$$

The matrices $A(s)$ and $B_c(s)$ being left-coprime, the polynomial controller form is obtained by solving the Bezout identities (8.39), and, more precisely, by finding polynomial matrices $E_0(s)$, $F_0(s)$, $A_1(s)$ and $B_1(s)$ satisfying:

$$\begin{cases} A(s)E_0(s) + B_c(s)F_0(s) = I \\ A(s)B_1(s) = B_c(s)A_1(s) \ . \end{cases}$$

This gives

$$\begin{cases} A_1(s) = s^2(s^2 - b) \\ B_1(s) = \begin{pmatrix} \beta_z(s) \\ \beta_\theta \end{pmatrix} = \begin{pmatrix} r_1 s^2 + ar_2 - br_1 \\ r_2 \end{pmatrix}, \end{cases} \tag{8.58}$$

and

$$\begin{cases} E_0(s) = \begin{pmatrix} \dfrac{r_1}{br_1 - ar_2} & \dfrac{r_1^2}{r_2(ar_2 - br_1)} \\ \dfrac{r_2}{br_1 - ar_2} & \dfrac{r_1}{ar_2 - br_1} \end{pmatrix} \\ F_0(s) = \begin{pmatrix} -\dfrac{s^2 - b}{br_1 - ar_2} & \dfrac{ar_2 - r_1 s^2}{r_2(ar_2 - br_1)} \end{pmatrix}. \end{cases} \tag{8.59}$$

8.10.2 Computation of the Closed-Loop Transfer Functions

The more general controller is of the form

$$p(s)u + q_z(s)y_1 + q_\theta(s)y_2 = r \ , \tag{8.60}$$

where r is a reference signal and q_z, q_θ are polynomials of degrees strictly less that the one of $p(s)$, for the controller to be causal. From a practical point of view, it is important to be able to act on the closed-loop behavior of the real physical variables of the systems z and θ, and not on their measurements y_1 and y_2 which are subjected to noises. If we denote by $\overline{y}$ the vector $(z, \theta)^\top$, then from (8.50) and the expression of the outputs

$$\begin{cases} y_1 = z + bz \\ y_2 = \theta + b\theta \ , \end{cases} \tag{8.61}$$

we obtain the different closed-loop transfers from w, d_m and r to $\overline{y}$:

$$\overline{y} = B_1T^{-1}r + (B_1T^{-1}(PF_0 - QE_0) + E_0)B_d w - B_1T^{-1}Qv \ . \tag{8.62}$$

Let us point out that, since u is scalar, $T(s)$ is a polynomial and not a polynomial matrix. The following transfer functions can then be written as

$$\begin{cases} T_{r\overline{y}}(s) = \frac{1}{T(s)} \begin{pmatrix} \beta_z(s) \\ \beta_\theta(s) \end{pmatrix} \\ T_{d_m\overline{y}}(s) = \dfrac{1}{T(s)} \begin{pmatrix} -q_z(s)\beta_z(s) & -q_\theta(s)\beta_z(s) \\ -q_z(s)\beta_\theta(s) & -q_\theta(s)\beta_\theta(s) \end{pmatrix} \\ T_{w\overline{y}}(s) = \dfrac{1}{T(s)} \begin{pmatrix} t_{11}(s) & t_{12}(s) \\ t_{21}(s) & t_{22}(s) \end{pmatrix}, \end{cases} \tag{8.63}$$

where the polynomials $t_{ij}(s)$ have the following form

$$\begin{cases} t_{11}(s) = p(s)h_1(s) + q_\theta(s)h_2(s) \\ t_{12}(s) = p(s)h_3(s) + q_\theta(s)h_4(s) \\ t_{21}(s) = p(s)h_5(s) + q_z(s)h_6(s) \\ t_{22}(s) = p(s)h_7(s) + q_z(s)h_8(s) \, , \end{cases} \tag{8.64}$$

the polynomials $h_i(s)$ being given by:

$$\begin{cases} h_1(s) = (r_1 A_1(s) - \beta_z(s)(s^2 - b))/(br_1 - ar_2) \\ h_2(s) = (r_1\beta_\theta(s) - r_2\beta_z(s))/(br_1 - ar_2) \\ h_3(s) = (r_1^2 A_1(s) + \beta_z(s)(ar_2 - r_1 s^2))/r_2(ar_2 - br_1) \\ h_4(s) = (r_1^2\beta_\theta(s) - r_1 r_2\beta_z(s))/r_2(ar_2 - br_1) \\ h_5(s) = (r_2 A_1(s) - \beta_\theta(s)(s^2 - b))/(br_1 - ar_2) \\ h_6(s) = (r_2\beta_z(s) - r_1\beta_\theta(s))/(br_1 - ar_2) \\ h_7(s) = (r_1 r_2 A_1(s) + \beta_\theta(s)(ar_2 - r_1 s^2))/r_2(ar_2 - br_1) \\ h_8(s) = (r_1 r_2\beta_z(s) - r_1^2\beta_\theta(s))/r_2(ar_2 - br_1) \, . \end{cases} \tag{8.65}$$

8.10.3 Affine Parameterization of the Controller

Let us now parameterize the controller (8.60) to use the degrees of freedom. The closed-loop poles given by the roots of the polynomial $T(s)$ being placed in the stability half-plane (see Definition 4.6 and Remark 8.44), polynomials $p(s)$, $q_z(s)$ and $q_\theta(s)$ have to be found satisfying:

$$p(s)A_1(s) + q_z(s)\beta_z(s) + q_\theta(s)\beta_\theta(s) = T(s) \, . \tag{8.66}$$

If p_0, q_{z_0} and q_{θ_0} constitute a particular solution, one can write:

$$A_1(p_0 - p) = \beta_z(q_z - q_{z0}) + \beta_\theta(q_\theta - q_{\theta_0}) \, . \tag{8.67}$$

Since the polynomials $\beta_z(s)$ and $\beta_\theta(s)$ given by (8.58) are coprime, we know by Corollary 8.21 that there exist polynomials r_z and r_θ satisfying:

$$r_z\beta_z + r_\theta\beta_\theta = -A_1 \, . \tag{8.68}$$

A particular solution is given by

$$r_z(s) = \frac{1}{r_1}s^2 \quad \text{and} \quad r_\theta(s) = -\frac{a}{r_1}\ , \tag{8.69}$$

with $\deg(r_z) = 2$ and $\deg(r_\theta) = 0$. Multiplying (8.68) by $(p - p_0)$, we obtain:

$$A_1(p_0 - p) = r_z\beta_z(p - p_0) + r_\theta\beta_\theta(p - p_0)\ . \tag{8.70}$$

Equations (8.67) and (8.70) are Bezout relations of type (8.13) with solutions $(q_z - q_{z0})$ and $(q_\theta - q_{\theta_0})$, on the one hand, and $r_z(p - p_0)$ and $r_\theta(p - p_0)$, on the other hand. Consequently, there exist polynomials k and l such that:

$$\begin{cases} p = p_0 + l \\ q_z = q_{z0} + lr_z + k\beta_\theta \\ q_\theta = q_{\theta_0} + lr_\theta - k\beta_z\ . \end{cases} \tag{8.71}$$

To obtain a particular solution, the division algorithm can be applied:

$$T = A_1 p_0 + r_0, \quad \deg(r_0) < \deg(A_1) = 4\ . \tag{8.72}$$

One has then to solve

$$r_0 = \beta_z q_{z0} + \beta_\theta q_{\theta_0}\ . \tag{8.73}$$

The unique solution must satisfy:

$$\begin{cases} \deg\left(q_{z0}(s)\right) < \deg\left(\beta_\theta(s)\right) = 2 \\ \deg\left(q_{\theta_0}(s)\right) < \deg\left(\beta_z(s)\right) = 2\ . \end{cases} \tag{8.74}$$

Then, from (8.69), (8.71) and (8.72), we have, with $\overline{n} = \deg\left(T(s)\right)$, that:

$$\begin{cases} \deg(q_z) = \max(\deg(l) + 2, \deg(k) + 2, 1) \\ \deg(q_\theta) = \max(\deg(l), \deg(k) + 2, 1) \\ \deg(p) = \max\left(\deg(p_0), \deg(l)\right) = \max\left(\overline{n} - 4, \deg(l)\right)\ . \end{cases} \tag{8.75}$$

But the polynomials l and k have also to respect the strict causality constraints of the controller, given by

$$\begin{cases} \deg(p) > \deg(q_z) \\ \deg(p) > \deg(q_\theta)\ . \end{cases} \tag{8.76}$$

If the polynomials l and k are chosen independently, the following inequalities can be deduced from (8.75) and (8.76):

$$\begin{cases} \deg(l) \leq \overline{n} - 7 \\ \deg(k) \leq \overline{n} - 7\ . \end{cases} \tag{8.77}$$

So, if $\overline{n} = 8$ poles are placed as expected, we obtain polynomials l and k of degree1 from (8.77), and therefore the 4 expected degrees of freedom: l^0, l^1, k^0, k^1 (coefficients of s^i in the corresponding polynomials).

8.10.4 *Placement of Regulation Zeros with Fixed Poles*

We are now going to use the degrees of freedom to asymptotically reject some classes of disturbances.

Asymptotic Rejection of Constant Perturbations

From (8.63)–(8.64), for the static gains (see Definition 3.33) from the dynamics noises w_1 and w_2 to the position z of the cart to be zero, it is sufficient to have:

$$p(0) = 0 \quad \text{and} \quad q_\theta(0) = 0\,. \tag{8.78}$$

In that case, the static gain from $b\theta$ to z is also zero but the one from bz to z is equal to -1, which means that a measurement bias on the position z cannot be rejected on the asymptotic value of z. If $q_z(0)$ would be zero, all the static gains from the w_i to θ would also be zero, but, from (8.66) and (8.78), we have that:

$$T(0) = q_z(0)\beta_z(0)\,, \quad \beta_z(0) \neq 0\,. \tag{8.79}$$

Consequently, $q_z(0)$ must be nonzero to guarantee the asymptotic stability of the closed-loop system. On the other hand, the fact that $\beta_\theta(0)$ is zero implies that the static gains from bz and $b\theta$ to θ are naturally zero. It means that without dynamics disturbance, the only possible equilibrium of the pendulum is the vertical position, corresponding to $\theta = 0$. Let us now summarize the conditions of *asymptotic rejection of constant perturbations*, denoted ARCP:

$$(\text{ARCP}) \qquad \begin{cases} p(0) = 0 \\ q_\theta(0) = 0 \\ q_z(0) \neq 0\,. \end{cases} \tag{8.80}$$

Using the general parameterization (8.71), these ARCP are equivalent to choosing

$$l^0 = -p_0^0 \quad \text{and} \quad k^0 = \frac{q_{\theta_0}^0 - p_0^0 r_\theta^0}{\beta_z^0}\,. \tag{8.81}$$

We can check afterwards that $q_{z0}(0)$ is not zero. Two degrees of freedom have been used to realize these ARCP conditions.

Asymptotic Rejection of Ramp Perturbations

From a dynamical point of view, a slope of the bench introduces a disturbance w_1 on the only carts's dynamics. When satisfying ARCP conditions (8.81), one is able to asymptotically reject on z a constant slope of the bench.

Remark 8.48 One could imagine that this slope is introduced regularly as a ramp by the public with a button driving a pneumatic jack. This is actually what was done on the full-size realization of the *double inverted pendulum* (*Cité des sciences et de l'industrie de la Villette*). Using the polynomial techniques developed in this chapter, we have elaborated a stabilizing control law which can, moreover, reject non measured disturbances, without increasing the size of the controller (see [25, 26]). This experiment had to deal with sensors and actuator noises, dry friction along the bench and the public could introduce a slope γ by driving the pneumatic jack at a constant velocity. ◇

This slope γ can then be described as a ramp disturbance (see Definition 3.11), namely,

$$w_1 = \frac{g\gamma}{s^2}\,, \tag{8.82}$$

where $g = 9.81\,\mathrm{m\,s^{-2}}$ is the constant of gravity. Using (8.63) and denoting $F_{\gamma z}(s)$ the transfer function from γ to z, we obtain

$$F_{\gamma z}(s) = \frac{t_{11}(s)}{T(s)} = \frac{p(s)h_1(s) + q_\theta(s)h_2(s)}{T(s)}\,,$$

where $h_1(s)$ and $h_2(s)$ are given by (8.64) and (8.65). If the ARCP conditions (8.81) hold true, $p(0)$ and $q_\theta(0)$ are zero, and it can be deduced that

$$t_{11}(0) = 0 = F_{\gamma z}(0)\,.$$

To asymptotically reject the ramp (8.82), the following relation between derivatives should also be satisfied (the symbol $'$ denotes the differentiation operation with respect to the variable s)

$$F'_{\gamma z}(0) = \frac{t'_{11}(0)T(0) - t_{11}(0)T'(0)}{T^2(0)} = 0\,.$$

Moreover, $t_{11}(0)$ being zero, one has to solve $t'_{11}(0) = 0 = p'(0)h_1(0) + q'_\theta(0)h_2(0)$. A solution can be obtained by ensuring:

$$p'(0) = q'_\theta(0) = 0\,. \tag{8.83}$$

Noticing that β_z^1 and r_θ^1 being zero, Eq. (8.83) implies the following choice for l^1 and k^1:

$$l^1 = -p_0^1 \quad \text{and} \quad k^1 = \frac{q_{\theta_0}^1 - p_0^1 r_\theta^0}{\beta_z^0} . \tag{8.84}$$

The two last degrees of freedom are then fixed to solve the asymptotic rejection of ramp perturbations.

To conclude, the polynomial approach makes it possible to easily parameterize a controller to make the degrees of freedom appear after the poles placement in the case of a multivariable system. These degrees of freedom can be used to face some disturbances. On the full-size realization of the double inverted pendulum, such stabilizing controllers have been implemented in real time and the results we have obtained have illustrated the importance of the choice of the regulation zeros.

8.11 Exercises

Exercise 8.11.1 Consider the following scalar l.c.s. system in state-space canonical controller form

$$\begin{cases} \dot{x} = Ax + Bu \\ y = Cx \end{cases} \tag{8.85}$$

with A, B and C of the form:

$$A = \begin{pmatrix} 0 & 1 & 0 \cdots & 0 & 0 \\ 0 & 0 & 1 \cdots & 0 & 0 \\ \cdot & \cdot & \cdots\cdot & \cdot & \cdot \\ \cdot & \cdot & \cdots\cdot & 1 & 0 \\ 0 & 0 & \cdots\cdot & 0 & 1 \\ -a_n & -a_{n-1} & \cdots\cdot & -a_2 & -a_1 \end{pmatrix}, \quad B = \begin{pmatrix} 0 \\ 0 \\ \cdot \\ \cdot \\ 0 \\ 1 \end{pmatrix}, \quad C = \begin{pmatrix} c_1 & \cdots & c_n \end{pmatrix}.$$

Compute the polynomial controller form of this system.

Exercise 8.11.2 Consider the scalar l.c.s. system (8.85) in state-space observer canonical form:

$$A = \begin{pmatrix} 0\,0 \cdots\cdot 0 & -a_n \\ 1\,0 \cdots\cdot 0 & -a_{n-1} \\ 0\,1 \cdots\cdot\cdot & \cdot \\ \cdot\cdot\cdots\cdot\cdot & \cdot \\ 0\,0 \cdots 1\,0 & -a_2 \\ 0\,0 \cdots 0\,1 & -a_1 \end{pmatrix}, \quad B = \begin{pmatrix} b_1 \\ b_2 \\ \cdot \\ \cdot \\ b_{n-1} \\ b_n \end{pmatrix}, \quad C = \begin{pmatrix} 0\,0 \cdots 0\,1 \end{pmatrix}.$$

Compute the polynomial observer form of this system.

Exercise 8.11.3 Consider the following l.c.s. system with input u and output y in polynomial observer form:

$$a(s)y = b(s)u + w \quad \text{with} \quad \deg\big(a(s)\big) > \deg\big(b(s)\big) \; .$$

The term w is a disturbance with an asymptotically stable dynamics given by:

$$d_w(s)w = 0 \; .$$

Let y_c be the output reference whose asymptotically stable dynamics is given by:

$$d_c(s)y_c = 0 \; .$$

We want to elaborate a polynomial controller

$$p(s)u = -q(s)y + q_c(s)y_c$$

with causality conditions

$$\deg\big(p(s)\big) > \deg(q(s)\big) \text{ and } \deg\big(p(s)\big) > \deg\big(q_c(s)\big) \; , \tag{8.86}$$

such that the output y asymptotically tracks the reference y_c in spite of the disturbance w.

1. Show that the transfer functions $w \to y$ and $y_c \to y$ can be written as

$$y = \frac{p(s)}{a(s)p(s) + b(s)q(s)} w + \frac{b(s)q_c(s)}{a(s)p(s) + b(s)q(s)} y_c \; .$$

2. Compute the dynamics of the error $e = y - y_c$. Show that, for the error to asymptotically tend to zero, it is sufficient for the controller to satisfy:
   - $p(s)$ is a multiple of the SCM (smallest common multiple) of the polynomials d_w and d_c;
   - $(q_c(s) - q(s))$ is a multiple of $d_c(s)$;
   - $(a(s)p(s) + b(s)q(s))$ is an *asymptotically stable polynomial* (viz. with roots having a strictly negative real part in the continuous-time case).

   The first condition means that the dynamics $p(s)$ of the controller contains those of the disturbance and of the set-point. This result is known as the "internal model principle" (see for example [70]).
3. Show that the following choice of controller is convenient

$$\begin{cases} p(s) &= p_0(s)\mathrm{SCM}(d_w(s), d_c(s)) \\ q(s) &= q_0(s)\mathrm{SCM}(d_w(s), d_c(s)) \\ q_c(s) &= K(s)d_c(s) \; , \end{cases}$$

   where the polynomials $p_0(s)$ and $q_0(s)$ satisfy:

$$a(s)p_0(s) + b(s)q_0(s) = \Delta(s) \; ,$$

$\Delta(s)$ being an "asymptotically stable polynomial" with a degree equal to the one of $a(s)$. Check that the causality condition (8.86) of the controller is satisfied.

Exercise 8.11.4 We consider the following system, describing the dynamics of a material point of mass m, disturbed by a noise w:

$$m\ddot{z} = F + w \ .$$

The position z of the point is assumed to be measured.

1. Put this system in a polynomial form as in (8.11.4) and give the polynomial observer and controller forms.
2. We consider the equation of the controller

   $$p(s)u + q(s)y = r \ ,$$

   r being a reference signal. The polynomial matrices $p(s)$ and $q(s)$ have to be determined to guarantee the closed-loop stability and also to act on the disturbance w. To do this, show that the closed-loop transfer $T_{wy}(s)$ from w to y can be written as

   $$T_{wy}(s) = \frac{p(s)}{T(s)} \ ,$$

   where $T(s) = p(s)ms^2 + q(s)$ is the polynomial describing the closed-loop system dynamics.
3. The disturbance w being a constant noise, show that 5 poles have to be placed to reject asymptotically w in the transfer $T_{wy}(s)$, namely to realize $p(0) = 0$.
4. Show that if w is a periodic disturbance of frequency $1/2\pi$, 6 poles have to be placed to reject asymptotically w in the transfer $T_{wy}(s)$, namely to realize $p(i) = p(-i) = 0$.

Appendix A
The Discrete-Time Stationary Riccati Equation

We study the stationary version of the induction equation (7.3), namely the Riccati equation (7.9).

A *symmetric matrix* $\mathtt{S}$ satisfies $\mathtt{S} = \mathtt{S}^\top$. Recall that a symmetric matrix $\mathtt{S}$ is said to be *positive* (respectively, *definite positive*) if, for all nonzero vector $\mathtt{x}$, we have that $\mathtt{x}^\top \mathtt{S}\mathtt{x} \geq 0$ (respectively, $\mathtt{x}^\top \mathtt{S}\mathtt{x} > 0$). If $\mathtt{S}_1$ and $\mathtt{S}_2$ are two symmetric matrices, we denote $\mathtt{S}_1 \geq \mathtt{S}_2$ if $\mathtt{S}_1 - \mathtt{S}_2$ is a positive matrix, and $\mathtt{S}_1 > \mathtt{S}_2$ if $\mathtt{S}_1 - \mathtt{S}_2$ is a definite positive matrix. A symmetric matrix $\mathtt{S}$ is said to be *definite negative* if $-\mathtt{S}$ is definite positive.

Definition A.1 *Consider* $\mathtt{F}$, $\mathtt{G}$ *and* $\mathtt{H}$, *three matrices of respective sizes* $n \times n$, $n \times m$ *and* $p \times n$, *and the mapping*

$$\psi(\Pi) = \mathtt{H}^\top \mathtt{H} + \mathtt{F}^\top \Pi \mathtt{F} - \mathtt{F}^\top \Pi \mathtt{G}(\mathtt{R} + \mathtt{G}^\top \Pi \mathtt{G})^{-1} \mathtt{G}^\top \Pi \mathtt{F} \,, \tag{A.1}$$

that maps any symmetric $n \times n$ *matrix* Π *into a symmetric* $n \times n$ *matrix* $\psi(\Pi)$. *We call* algebraic (or stationary) discrete-time Riccati equation *the equation* $\Pi = \psi(\Pi)$, *namely*

$$\Pi = \mathtt{H}^\top \mathtt{H} + \mathtt{F}^\top \Pi \mathtt{F} - \mathtt{F}^\top \Pi \mathtt{G}(\mathtt{R} + \mathtt{G}^\top \Pi \mathtt{G})^{-1} \mathtt{G}^\top \Pi \mathtt{F} \,. \tag{A.2}$$

Let Δ *be a matrix of size* $n \times m$ *and* Π *an* $n \times n$ *symmetric matrix. We set*

$$\psi_\Delta(\Pi) = \mathtt{H}^\top \mathtt{H} + (\mathtt{F} - \mathtt{G}\Delta)^\top \Pi (\mathtt{F} - \mathtt{G}\Delta) + \Delta \mathtt{R} \Delta \tag{A.3a}$$

$$\bar{\Delta}(\Pi) = (\mathtt{G}^\top \Pi \mathtt{G} + R)^{-1} \mathtt{G}^\top \Pi \mathtt{F} \tag{A.3b}$$

$$\widetilde{\mathtt{F}}(\Pi) = \mathtt{F} - \mathtt{G}\bar{\Delta}(\Pi) = \mathtt{F} - \mathtt{G}(\mathtt{R} + \mathtt{G}^\top \Pi \mathtt{G})^{-1} \mathtt{G}^\top \Pi \mathtt{F} \,. \tag{A.3c}$$

The following Lemma A.2 is proven in [14, 15].

Lemma A.2 ([14, 15]) *For all symmetric matrix* Π, *we have that*

$$\psi_{\bar{\Delta}(\Pi)}(\Pi) = \psi(\Pi) \,. \tag{A.4}$$

B. d'Andréa-Novel and M. De Lara, *Control Theory for Engineers*,
DOI: 10.1007/978-3-642-34324-7, © Springer-Verlag Berlin Heidelberg 2013

Moreover, the mapping ψ is minimal in the sense that

$$\Pi_1 \leq \Pi_2 \Rightarrow \psi(\Pi_1) \leq \inf_{\Delta} \psi_{\Delta}(\Pi_2) \leq \psi(\Pi_2) \ . \tag{A.5}$$

Lemma A.3 *There is a unique positive solution Π of the Riccati equation* (A.2) *such that the matrix $\widetilde{\mathtt{F}}(\Pi)$ is asymptotically stable.*

Proof By Proposition 7.4, uniqueness is straightforward since $\mathtt{x}_0^\top \Pi \mathtt{x}_0$ is *the* minimum (7.10) of the associated optimization problem. □

Proposition A.4 *If the couple* $(\mathtt{F}, \mathtt{G})$ *is controllable and if the couple* $(\mathtt{F}, \mathtt{H})$ *(or* $(\mathtt{F}, \mathtt{HF})$*) is observable, there exists a unique symmetric positive solution Π to the algebraic Riccati equation* (A.2) *such that the matrix $\widetilde{\mathtt{F}}(\Pi)$ is asymptotically stable. Moreover, this matrix Π is positive* definite *and is obtained as the limit of any sequence*

$$\Pi_{k+1} = \psi(\Pi_k) \ , \ k \in \mathbb{N} \ , \tag{A.6}$$

for any initial condition $\Pi_0 \geq 0$.

The proof consists of two lemmas.

Lemma A.5 *Consider the sequence $\left(\Pi_k^0\right)_{k \geq 0}$ of symmetric matrices defined by the induction*

$$\Pi_{k+1}^0 = \psi(\Pi_k^0) \ , \ k \in \mathbb{N} \ , \ \Pi_0^0 = 0 \ . \tag{A.7}$$

1. *The sequence $\left(\Pi_k^0\right)_{k \geq 0}$ is increasing, and made of symmetric positive matrices.*
2. *If the couple* $(\mathtt{F}, \mathtt{G})$ *is controllable, the sequence $\left(\Pi_k^0\right)_{k \geq 0}$ is bounded above and converges towards a symmetric positive matrix Π_∞.*
3. *If the couple* $(\mathtt{F}, \mathtt{H})$ *is observable, the sequence $\left(\Pi_k^0\right)_{k \geq n}$ is bounded below by $\Pi_n^0 > 0$.*
4. *If the couple* $(\mathtt{F}, \mathtt{HF})$ *is observable, the sequence $\left(\Pi_k^0\right)_{k \geq n+1}$ is bounded below by $\Pi_{n+1}^0 > 0$.*

Proof It is clear that the sequence $\left(\Pi_k^0\right)_{k \geq 0}$ is made of symmetric matrices (and positive if we prove that the sequence is increasing).

1. Let us consider the sequence of criteria (without final state penalization)

$$J_f(\mathtt{u}) = J(\mathtt{u}_0, \ldots, \mathtt{u}_{f-1}) = \sum_{k=0}^{f-1} \left(\mathtt{u}_k^\top \mathtt{R} \mathtt{u}_k + \mathtt{x}_k^\top \mathtt{H}^\top \mathtt{H} \mathtt{x}_k \right) \ , \ f \geq 1 \ ,$$

where $\mathtt{x}_k$ solves the linear dynamical system

$$\mathtt{x}_{k+1} = \mathtt{F}\mathtt{x}_k + \mathtt{G}\mathtt{u}_k \ , \ k \in \mathbb{N} \ . \tag{A.8}$$

As these criteria are increasing in f, the same property holds true for their minimal values and thus, by Proposition 7.2 and 7.4, we have that:

$$\mathrm{x}_0^\top \Pi_{f+1}^0 \mathrm{x}_0 \geq \mathrm{x}_0^\top \Pi_f^0 \mathrm{x}_0 \geq \cdots \geq \mathrm{x}_0^\top \Pi_1^0 \mathrm{x}_0 \geq 0 \, .$$

2. If the couple (F, G) is controllable, there exists, by Theorem 6.23, a matrix K such that $(\mathrm{F}_d - \mathrm{G}_d \mathrm{K})$ is asymptotically stable. Taking for control sequence u the one given by the state feedback control (6.31), we have that

$$J(\mathrm{u}) = \sum_{k=0}^{+\infty} \left(\mathrm{u}_k^\top \mathrm{R} \mathrm{u}_k + \mathrm{x}_k^\top \mathrm{Q} \mathrm{x}_k \right) < +\infty \, ,$$

since x_k and u_k decrease exponentialy towards 0 (see the proof of Proposition 6.12). This yields the inequality

$$\mathrm{x}_0^\top \Pi_f^0 \mathrm{x}_0 \leq J_f(\mathrm{u}) \leq J(\mathrm{u}) < +\infty \, .$$

The sequence $\left(\mathrm{x}_0^\top \Pi_f^0 \mathrm{x}_0 \right)_{f \geq 0}$, increasing and bounded above for each x_0, converges thus in $\mathbb{R}$. We easily deduce that the sequence of symmetric positive matrices $\left(\Pi_f^0 \right)_{f \geq 0}$ converges towards the symmetric positive matrix Π_∞ characterized by the expression

$$\mathrm{x}_0^\top \Pi_\infty \mathrm{z}_0 = \frac{1}{2} \lim_{f \to +\infty} \left((\mathrm{x}_0 + \mathrm{z}_0)^\top \Pi_f^0 (\mathrm{x}_0 + \mathrm{z}_0) - \mathrm{x}_0^\top \Pi_f^0 \mathrm{x}_0 - \mathrm{z}_0^\top \Pi_f^0 \mathrm{z}_0 \right) \, .$$

3. If the couple (F, H) is observable, let us show that the symmetric matrix Π_n^0 is not only positive, but also definite. Indeed, if x_0 is such that $\mathrm{x}_0^\top \Pi_n^0 \mathrm{x}_0 = 0$, then, with the notations of the proof of Proposition 7.4, we have that

$$0 = J_n(\mathrm{u}^\star) = \sum_{k=0}^{n-1} \left(\mathrm{u}_k^{\star\top} \mathrm{R} \mathrm{u}_k^\star + \mathrm{x}_k^{\star\top} \mathrm{H}^\top \mathrm{H} \mathrm{x}_k^\star \right) \, .$$

We deduce that $\mathrm{u}_k^{\star\top} \mathrm{R} \mathrm{u}_k^\star = 0$ and $\mathrm{x}_k^{\star\top} \mathrm{H}^\top \mathrm{H} \mathrm{x}_k^\star = \|\mathrm{H}\mathrm{x}_k\|^2 = 0$, namely $\mathrm{u}_0^\star = \cdots = \mathrm{u}_{n-1}^\star = 0$ and $\mathrm{H}\mathrm{x}_0^\star = \cdots = \mathrm{H}\mathrm{x}_{f-1}^\star = 0$. As the control is zero, the solution $(\mathrm{x}_k^\star)_{k \in \mathbb{N}}$ to (A.8) satisfies $\mathrm{x}_{k+1}^\star = \mathrm{F}\mathrm{x}_k^\star$, and we obtain

$$H \mathrm{x}_0^\star = \mathrm{H}\mathrm{F}\mathrm{x}_0^\star = \cdots = \mathrm{H}\mathrm{F}^{n-1} \mathrm{x}_0^\star = 0 \, .$$

By observability of the couple $(\mathtt{F}, \mathtt{H})$, we deduce that $\mathtt{x}_0 = \mathtt{x}_0^\star = 0$ (see Corollary 6.26).

4. If the couple $(\mathtt{F}, \mathtt{HF})$ is observable, let us show that the symmetric matrix Π_{n+1}^0 is not only positive but also definite. Indeed, if $\mathtt{x}_0$ is such that $\mathtt{x}_0^\top \Pi_{n+1}^0 \mathtt{x}_0 = 0$, then, as above, we deduce that

$$\mathtt{HF}\mathtt{x}_0^\star = \cdots = \mathtt{HF}^{n-1}\mathtt{x}_0^\star = \mathtt{HF}^n\mathtt{x}_0^\star = 0 \,.$$

By observability of the couple $(\mathtt{F}, \mathtt{HF})$, we deduce that $\mathtt{x}_0 = \mathtt{x}_0^\star = 0$.

This concludes the proof. □

Lemma A.6 *If the couple* $(\mathtt{F}, \mathtt{G})$ *is controllable and if the couple* $(\mathtt{F}, \mathtt{H})$ *(or* $(\mathtt{F}, \mathtt{HF})$*) is observable, the symmetric positive matrix* Π_∞ *provided by item 2 in Lemma A.5 is such that*

1. *the matrix* Π_∞ *satisfies the stationary Riccati equation*

$$\Pi_\infty = \mathtt{H}^\top\mathtt{H} + \mathtt{F}^\top\Pi_\infty\mathtt{F} - \mathtt{F}^\top\Pi_\infty\mathtt{G}(R + \mathtt{G}^\top\Pi_\infty\mathtt{G})^{-1}\mathtt{G}^\top\Pi_\infty\mathtt{F} \quad ; \qquad \text{(A.9)}$$

2. *the matrix* $\widetilde{\mathtt{F}}(\Pi_\infty)$ *is asymptotically stable;*
3. *for all initial condition* $\Pi_0 \geq 0$*, the sequence (A.6) converges towards* Π_∞*.*

Proof The proof is inspired from [14, 15]. By Lemma A.5, taking the limit in (A.7) provides the equality (A.9).

We are going to show that, for k large enough, the matrix $\widetilde{\mathtt{F}}(\Pi_\infty)^k$ is asymptotically stable, which implies that $\widetilde{\mathtt{F}}(\Pi_\infty)$ is asymptotically stable (by comparing their eigenvalues). Since $\Pi_\infty > 0$ by Lemma A.5, it suffices to show that, for k large enough, the matrix $\widetilde{\mathtt{F}}\left(\Pi_\infty^\top\right)^k \Pi_\infty\widetilde{\mathtt{F}}\left(\Pi_\infty\right)^k - \Pi_\infty$ is negative definite by Propositions 6.16. Now, thanks to (A.3c) and to (A.4), the equality (A.9) can be written as

$$\Pi_\infty = \widetilde{\mathtt{F}}(\Pi_\infty)^\top\Pi_\infty\widetilde{\mathtt{F}}(\Pi_\infty) + \mathtt{C}^\top\mathtt{C} \,, \qquad \text{(A.10)}$$

where $\mathtt{C}$ denotes a square root with the following expression:

$$\mathtt{C}^\top\mathtt{C} = \mathtt{H}^\top\mathtt{H} + \bar{\Delta}(\Pi_\infty)\mathrm{R}\bar{\Delta}(\Pi_\infty) \,. \qquad \text{(A.11)}$$

Equation (A.10) successively applied yields

$$\Pi_\infty - \widetilde{\mathtt{F}}\left(\Pi_\infty^\top\right)^k \Pi_\infty\widetilde{\mathtt{F}}\left(\Pi_\infty\right)^k = \sum_{i=0}^{k-1}\widetilde{\mathtt{F}}\left(\Pi_\infty^\top\right)^i \mathtt{C}^\top\mathtt{C}\widetilde{\mathtt{F}}\left(\Pi_\infty\right)^i \,.$$

It suffices then to show that the right term is positive definite for k large enough. By Corollary 6.26, this is the case when the couple $(\widetilde{\mathtt{F}}(\Pi_\infty), \mathtt{C})$ (or $(\widetilde{\mathtt{F}}(\Pi_\infty), \mathtt{C}\widetilde{\mathtt{F}}(\Pi_\infty))$)

is observable. As a consequence, the couple $(\mathtt{F}, \mathtt{H})$ (or $(\mathtt{F}, \mathtt{HF})$) is observable. Indeed, suppose that there exists a vector $\mathtt{x}$ such that

$$\mathtt{Cx} = \mathtt{C}\widetilde{\mathtt{F}}(\Pi_\infty)\mathtt{x} = \cdots = \mathtt{C}\widetilde{\mathtt{F}}(\Pi_\infty)^k\mathtt{x} = 0 \, .$$

We deduce that

$$\begin{aligned} \mathtt{Cx} = 0 &\iff \mathtt{Hx} = 0 \text{ and } \bar{\Delta}(\Pi_\infty)\mathtt{x} = 0 \text{ by (A.11)} \\ &\Rightarrow \widetilde{\mathtt{F}}(\Pi_\infty)\mathtt{x} = \mathtt{Fx} \text{ by (A.3c).} \end{aligned}$$

Proceeding in the same way, we obtain

$$\begin{aligned} \mathtt{C}\widetilde{\mathtt{F}}(\Pi_\infty)\mathtt{x} = 0 &\iff \mathtt{CFx} = 0 \\ &\iff \mathtt{HFx} = 0 \text{ and } \bar{\Delta}(\Pi_\infty)\mathtt{Fx} = 0 \\ &\Rightarrow \widetilde{\mathtt{F}}(\Pi_\infty)^2\mathtt{x} = \mathtt{F}^2\mathtt{x} \, , \end{aligned}$$

and, step by step, also the following implication

$$\mathtt{Cx} = \mathtt{C}\widetilde{\mathtt{F}}(\Pi_\infty)\mathtt{x} = \cdots = \mathtt{C}\widetilde{\mathtt{F}}(\Pi_\infty)^k\mathtt{x} = 0 \Rightarrow \mathtt{Hx} = \mathtt{HFx} = \cdots = \mathtt{HF}^k\mathtt{x} = 0 \, .$$

Thus, if the couple $(\mathtt{F}, \mathtt{H})$ (respectively, $(\mathtt{F}, \mathtt{HF})$) is observable, then $\mathtt{x} = 0$ and the couple $(\widetilde{\mathtt{F}}(\Pi_\infty), \mathtt{C})$ (respectively, $(\widetilde{\mathtt{F}}(\Pi_\infty), \mathtt{C}\widetilde{\mathtt{F}}(\Pi_\infty))$) is also observable by Corollary 6.26.

Finally, we show the last item. By the inequality (A.5), the sequence (A.6) satisfies

$$\Pi_{k+1} = \psi(\Pi_k) \leq \psi_{\bar{\Delta}(\Pi_\infty)}(\Pi_k) = \mathtt{H}^\top\mathtt{H} + \widetilde{\mathtt{F}}(\Pi_\infty)^\top \Pi_k \widetilde{\mathtt{F}}(\Pi_\infty) + \bar{\Delta}(\Pi_\infty)\mathtt{R}\bar{\Delta}(\Pi_\infty)\bar{\Delta} \, .$$

By substracting the equality (A.9) from this inequality, we find that

$$\Pi_{k+1} - \Pi_\infty \leq \widetilde{\mathtt{F}}(\Pi_\infty)^\top(\Pi_k - \Pi_\infty)\widetilde{\mathtt{F}}(\Pi_\infty) \, ,$$

and thus:

$$\Pi_k - \Pi_\infty \leq \widetilde{\mathtt{F}}\left(\Pi_\infty^\top\right)^k (\Pi_0 - \Pi_\infty)\widetilde{\mathtt{F}}(\Pi_\infty)^k \, .$$

For all vector $\mathtt{x}_0$, by taking the limit, we obtain

$$\limsup_{k \to +\infty} \mathtt{x}_0^\top \Pi_k \mathtt{x}_0 \leq \mathtt{x}_0^\top \Pi_\infty \mathtt{x}_0 \tag{A.12}$$

since $\widetilde{\mathtt{F}}(\Pi_\infty)$ is asymptotically stable. On the other hand, by the inequality (A.5), the sequences given by the inductions (A.6) and (A.7) satisfy

$$\Pi_k^0 \leq \Pi_k \, , \quad \forall k \in \mathbb{N} \, ,$$

since $\Pi_0^0 = 0 \leq \Pi_0$. For all vector x_0, by taking the limit, we obtain

$$x_0^\top \Pi_\infty x_0 = \lim_{k \to +\infty} x_0^\top \Pi_k^0 x_0 \leq \liminf_{k \to +\infty} x_0^\top \Pi_k x_0 \ . \tag{A.13}$$

From the inequalities (A.12) and (A.13), we deduce that Π_k converges towards Π_∞. □

Appendix B
Laplace Transform and z-Transform

We provide some results on Laplace transform and z-transform. We refer the reader to [60, Chap. 6] for the proofs and also for details about the Laplace transform of distributions.

B.1 Laplace Transform

The Laplace transform maps any locally integrable function $\mathtt{f}$ of the real variable t, zero for $t < 0$ and satisfying some appropriate restrictive conditions, onto the function of the complex variable $s \in \mathbb{C}$ defined by:

$$\mathcal{L}[\mathtt{f}](s) := \int_0^{+\infty} e^{-st}\mathtt{f}(t)\,\mathrm{d}t \; . \tag{B.1}$$

Lemma B.1 *Let* $\mathtt{f}$ *be a locally integrable function defined on* $[0, +\infty[$. *There exists an extended real* $a \in \mathbb{R} \cup \{-\infty, +\infty\}$ *such that, for all* $s \in \mathbb{C}$,

- *if* $\Re(s) > a$, *then* $\int_0^{+\infty} e^{-st}|\mathtt{f}(t)|\,\mathrm{d}t < +\infty$;
- *if* $\Re(s) < a$, *then* $\int_0^{+\infty} e^{-st}|\mathtt{f}(t)|\,\mathrm{d}t = +\infty$.

The extended real a *is called the* abscissa of convergence *and the domain* $\{s \in \mathbb{C} \mid \Re(s) > a\}$ *the* region of convergence.

Definition B.2 *We call* Laplace transform *of the locally integrable function* $\mathtt{f}$ *the function* $\mathcal{L}[\mathtt{f}]$ *of the complex variable* s *defined by (B.1) on the region of convergence. The complex variable* s *is called the* Laplace variable.

Remark B.3 This definition can be extended to distributions as follows. Let $\mathtt{T}$ be a distribution on the real axis of the time variable t, with a support included in $[0, +\infty[$.

B. d'Andréa-Novel and M. De Lara, *Control Theory for Engineers*,
DOI: 10.1007/978-3-642-34324-7, © Springer-Verlag Berlin Heidelberg 2013

If there exists a scalar ξ_0 such that, for $\xi > \xi_0$, $e^{-\xi t}\mathtt{T}$ is a tempered distribution, then its Laplace transform can be defined as follows:

$$\mathcal{L}\left[\mathtt{T}\right](s) = \left\langle \mathtt{T}, e^{-st} \right\rangle \text{ for } \Re(s) > \xi_0 . \tag{B.2}$$

◇

Example B.4 $\mathcal{L}\left[\delta\right](s) = 1, \quad \mathcal{L}\left[\delta^{(m)}\right](s) = s^m, \quad \mathcal{L}\left[\delta_a\right](s) = e^{-as}.$ △

We now give some important properties of the Laplace transform. Since the Laplace transform is only defined on a region of convergence, one should pay attention to these regions of convergence when stating the different properties. We show how to proceed for the first property and then let the reader complete for the other ones.

L1. The Laplace transform is linear, namely: if $\mathtt{f}_1$ and $\mathtt{f}_2$ are two locally integrable functions defined on $[0, +\infty[$ and λ is a real number, the Laplace transform of the function $\lambda\mathtt{f}_1+\mathtt{f}_2$ exists on the intersection of the two regions of convergence and satisfies
$$\mathcal{L}\left[\lambda\mathtt{f}_1 + \mathtt{f}_2\right](s) = \lambda\mathcal{L}\left[\mathtt{f}_1\right](s) + \mathcal{L}\left[\mathtt{f}_2\right](s) .$$

L2. If $\mathtt{H}(s)$ denotes the Laplace transform of $\mathtt{h}(t)$ and $\mathtt{U}(s)$ that of $\mathtt{u}(t)$, then the Laplace transform of the *convolution product* $\mathtt{h} \star \mathtt{u}$ is given by
$$\mathcal{L}\left[\mathtt{h} \star \mathtt{u}\right](s) = \mathtt{H}(s)\mathtt{U}(s) , \tag{B.3}$$
where we recall that
$$(\mathtt{h} \star \mathtt{u})(t) = \int_0^t \mathtt{h}(\tau)\mathtt{u}(t-\tau)\,\mathrm{d}\tau = \int_0^t \mathtt{h}(t-\tau)\mathtt{u}(\tau)\,\mathrm{d}\tau . \tag{B.4}$$

L3. Time shifting Theorem. Consider a locally integrable function $\mathtt{f}$. Let us denote $\Theta_a\mathtt{f}(t) := \mathtt{f}(t-a)$, for any $a \in \mathbb{R}$. We have that
$$\mathcal{L}\left[\Theta_a\mathtt{f}\right](s) = e^{-as}\mathcal{L}\left[\mathtt{f}\right](s) . \tag{B.5}$$

L4. Differentiation Theorem. Consider a function $\mathtt{f}$ having derivatives up to order m. If $\mathtt{f}^{(m)}$ is locally integrable, then:
$$\mathcal{L}\left[\mathtt{f}^{(m)}\right](s) = s^m\mathcal{L}\left[\mathtt{f}\right](s) - s^{m-1}\mathtt{f}(0) - \cdots - \mathtt{f}^{(m-1)}(0) . \tag{B.6}$$

L5. Integration Theorem. For a locally integrable function $\mathtt{f}$, we have that
$$\mathcal{L}\left[\int_0^{\cdot} \mathtt{f}(\tau)\,\mathrm{d}\tau\right](s) = \frac{1}{s}\mathcal{L}\left[\mathtt{f}\right](s) . \tag{B.7}$$

Table B.1 Laplace transforms

$\mathtt{f}(t)$	$\mathcal{L}[\mathtt{f}](s)$	Region of convergence
δ	1	$\mathbb{C}$
δ_a	e^{-sa}	$\mathbb{C}$
$\mathfrak{E}(t)$	$\frac{1}{s}$	$\{s \in \mathbb{C} \mid \Re(s) > 0\}$
$\frac{t^{m-1}}{(m-1)!}\mathfrak{E}(t)$	$\frac{1}{s^m}$	$\{s \in \mathbb{C} \mid \Re(s) > 0\}$
$e^{at}\frac{t^{m-1}}{(m-1)!}\mathfrak{E}(t)$	$\frac{1}{(s-a)^m}$	$\{s \in \mathbb{C} \mid \Re(s) > a\}$
$e^{at}\sin(bt)\mathfrak{E}(t)$	$\frac{b}{(s-a)^2+b^2}$	$\{s \in \mathbb{C} \mid \Re(s) > a\}$
$e^{at}\cos(bt)\mathfrak{E}(t)$	$\frac{s-a}{(s-a)^2+b^2}$	$\{s \in \mathbb{C} \mid \Re(s) > a\}$

L6. Initial value Theorem. Consider a locally integrable function $\mathtt{f}$. If the following limits exist, they are equal:

$$\lim_{t\to 0} \mathtt{f}(t) = \lim_{s\to+\infty} s\mathcal{L}[\mathtt{f}](s) . \tag{B.8}$$

L7. Final value Theorem. Consider a locally integrable function $\mathtt{f}$. If the following limits exist, they are equal:

$$\lim_{t\to+\infty} \mathtt{f}(t) = \lim_{s\to 0} s\mathcal{L}[\mathtt{f}](s) . \tag{B.9}$$

The Table B.1 provides a list of usual Laplace transforms.

Here, δ_a denotes the Dirac delta function at point a ($\delta_0 = \delta$) and $\mathfrak{E}(t)$ the Heaviside step function introduced in Definition 3.11 ($\mathfrak{E}(t) = 1$ if $t \geq 0$, and $\mathfrak{E}(t) = 0$ else).

B.2 The z-Transform

The discrete-time counterpart of the Laplace transform of locally integrable functions on $[0, +\infty[$ is the z-transform of sequences indexed by $\mathbb{N}$.

Definition B.5 *We call* z-transform *of the sequence* $\mathtt{x} = (\mathtt{x}_k)_{k\in\mathbb{N}}$ *the power series defined by*

$$\mathcal{Z}[\mathtt{x}](z) := \sum_{k=0}^{+\infty} \mathtt{x}_k z^{-k} . \tag{B.10}$$

This series is absolutely convergent if $|z| > R$*, where* R *denotes the* radius of convergence.

Let us now recall some important properties of the z-transform. The precautions taken in continuous-time remain valid here. When not specified, $\mathtt{x} = (\mathtt{x}_k)_{k\in\mathbb{N}}$ denotes a real-valued sequence.

Z1. The z-transform is linear.

Z2. If we denote $\mathtt{H}(z)$ the z-transform of the sequence $(\mathtt{h}_k)_{k\in\mathbb{N}}$ and $\mathtt{U}(z)$ that of the sequence $(\mathtt{u}_k)_{k\in\mathbb{N}}$, then the z-transform of the *convolution product* $\mathtt{h}\star\mathtt{u}$ is given by

$$\mathcal{Z}\left[\mathtt{h}\star\mathtt{u}\right](z) = \mathtt{H}(z)\mathtt{U}(z) \ , \tag{B.11}$$

where we recall that

$$(\mathtt{h}\star\mathtt{u})_k = \sum_l \mathtt{h}_l \mathtt{u}_{k-l} = \sum_l \mathtt{h}_{k-l}\mathtt{u}_l \ . \tag{B.12}$$

Z3. Time shifting Theorem. If, for any $l \in \mathbb{N}$, the delay operator Θ_l is defined by $(\Theta_l \mathtt{x})_n := \mathtt{x}_{n-l}$, then

$$\mathcal{Z}\left[\Theta_l \mathtt{x}\right] = z^{-l}\mathcal{Z}\left[\mathtt{x}\right](z) \ . \tag{B.13}$$

Z4. Advance Theorem. For any $l \in \mathbb{N}$, we have that:

$$\mathcal{Z}\left[\Theta_{-l}\mathtt{x}\right] = z^l\left(\mathcal{Z}\left[\mathtt{x}\right](z) - \sum_{i=0}^{l-1}\mathtt{x}_i z^{-i}\right) \ . \tag{B.14}$$

Z5. Initial value Theorem. If the limit exists, we have that:

$$\mathtt{x}_0 = \lim_{z\to+\infty}\mathcal{Z}\left[\mathtt{x}\right](z) \ . \tag{B.15}$$

Z6. Final value Theorem. When the following limits exist, they are equal:

$$\lim_{k\to+\infty}\mathtt{x}_k = \lim_{z\to 1}(1 - z^{-1})\mathcal{Z}\left[\mathtt{x}\right](z) \ . \tag{B.16}$$

The Table B.2 gives a list of usual z-transforms.

Here, $\delta_{i,j}$ denotes the Kronecker symbol ($\delta_{i,j} = 0$ if $i \neq j$, 1 else) and $\mathfrak{E}_n$ denotes the *unit-step sequence* ($\mathfrak{E}_n = 1$ if $n \geq 0$ and $\mathfrak{E}_n = 0$ if $n < 0$).

Table B.2 Usual z-transforms

x_n	$\mathcal{Z}[x](z)$	Radius of convergence
$\delta_{n,0}$	1	$+\infty$
$\delta_{n,k}$	z^{-k}	$+\infty$
$\mathfrak{E}_n$	$\dfrac{1}{1-z^{-1}}$	1
$nT\,\mathfrak{E}_n$	$\dfrac{Tz^{-1}}{(1-z^{-1})^2}$	1
$e^{-anT}\,\mathfrak{E}_n$	$\dfrac{1}{1-e^{-aT}z^{-1}}$	e^{-aT}
$r^n\,\mathfrak{E}_n$	$\dfrac{1}{1-rz^{-1}}$	r
$r^n \sin(bn)\mathfrak{E}_n$	$\dfrac{(\sin b)z^{-1}}{1-2(\cos b)rz^{-1}+r^2z^{-2}}$	r
$r^n \cos(bn)\mathfrak{E}_n$	$\dfrac{1-(\cos b)z^{-1}}{1-2(\cos b)rz^{-1}+r^2z^{-2}}$	r

Appendix C
Gaussian Vectors

We suppose that the reader is familiar with basic notions of probability calculus: random variable, expectation, variance, dispersion, conditional expectation, independence. We refer the reader to [11, 31].

C.1 Recalls of Probability Calculus

Let $(\Omega, \mathcal{F}, \mathbb{P})$ be a probability space. The *mathematical expectation* with respect to the probability $\mathbb{P}$ is denoted by $\mathbb{E}$.

Definition C.1 *The* characteristic function *of a random vector* $\mathtt{Z} = (\mathtt{Z}_1, \ldots, \mathtt{Z}_r)$ *of dimension* r *is the following mapping* $\Phi_\mathtt{Z} : \mathbb{R}^r \to \mathbb{C}$*:*

$$\Phi_\mathtt{Z}(\theta) = \mathbb{E}\big[\exp(i\theta^\top \mathtt{Z})\big] = \mathbb{E}\big[(\exp(i\theta_1 \mathtt{Z}_1 + \cdots + i\theta_r \mathtt{Z}_r)\big] \ , \ \ \forall \theta \in \mathbb{R}^r \ . \tag{C.1}$$

Proposition C.2 *Two random vectors have the same distribution if, and only if, they have the same characteristic function. Two random vectors* $\mathtt{X}$ *and* $\mathtt{Y}$ *(with respective dimensions* $r_\mathtt{X}$ *and* $r_\mathtt{Y}$*) are independent if, and only if,*

$$\Phi_{(\mathtt{X},\mathtt{Y})}(\theta_\mathtt{X}, \theta_\mathtt{Y}) = \Phi_\mathtt{X}(\theta_\mathtt{X})\Phi_\mathtt{Y}(\theta_\mathtt{Y}) \ , \ \ \forall (\theta_\mathtt{X}, \theta_\mathtt{Y}) \in \mathbb{R}^{r_\mathtt{X}} \times \mathbb{R}^{r_\mathtt{Y}} \ . \tag{C.2}$$

A real-valued random variable $\mathtt{X}$ is said to be *square integrable* if $\mathtt{X}^2$ is integrable, that is, $\mathbb{E}(\mathtt{X}^2) < +\infty$. In that case, the mean

$$\overline{\mathtt{X}} := \mathbb{E}[\mathtt{X}] \tag{C.3}$$

exists, and we call *variance* the quantity

B. d'Andréa-Novel and M. De Lara, *Control Theory for Engineers*,
DOI: 10.1007/978-3-642-34324-7, © Springer-Verlag Berlin Heidelberg 2013

$$\mathsf{var}(\mathrm{X}) := \mathbb{E}\big[(\mathrm{X} - \overline{\mathrm{X}})^2\big] = \mathbb{E}(\mathrm{X}^2) - \mathbb{E}(\mathrm{X})^2 \,. \tag{C.4}$$

A random vector X is said to be *square integrable* if the norm $\|\mathrm{X}\|$ is square integrable, that is, if $\mathbb{E}(\|\mathrm{X}\|^2) < +\infty$; it is said to have moments of all orders if $\mathbb{E}(e^{a\|\mathrm{X}\|}) < +\infty$, for all $a > 0$.

Definition C.3 *Let* X *and* Y *be two square integrable random vectors (with respective dimensions* r_{X} *and* r_{Y}*). We call* covariance *of* X *and* Y *the matrix of dimension* $r_{\mathrm{X}} \times r_{\mathrm{Y}}$ *defined by:*

$$\mathsf{cov}(\mathrm{X}, \mathrm{Y}) := \mathbb{E}\big[(\mathrm{X} - \overline{\mathrm{X}})(\mathrm{Y} - \overline{\mathrm{Y}})^\top\big] = \mathbb{E}(\mathrm{X}\mathrm{Y}^\top) - \mathbb{E}(\mathrm{X})\mathbb{E}(\mathrm{Y}^\top) \,. \tag{C.5}$$

We say that two random vectors are decorrelated *if their covariance is zero.*

We call dispersion *of* X, *or* covariance matrix, *or* variance-covariance matrix, *and we note* $\mathsf{D}(\mathrm{X})$ *the symmetric matrix*

$$\mathsf{D}(\mathrm{X}) := \mathsf{cov}(\mathrm{X}, \mathrm{X}) = \mathbb{E}\big[(\mathrm{X} - \overline{\mathrm{X}})(\mathrm{X} - \overline{\mathrm{X}})^\top\big] \,. \tag{C.6}$$

Remark C.4 If X and Y are independent, they are decorrelated since

$$\mathsf{cov}(\mathrm{X}, \mathrm{Y}) = \mathbb{E}\big[(\mathrm{X} - \overline{\mathrm{X}})(\mathrm{Y} - \overline{\mathrm{Y}})^\top\big] = \mathbb{E}(\mathrm{X} - \overline{\mathrm{X}})\mathbb{E}\big[(\mathrm{Y} - \overline{\mathrm{Y}})^\top\big] = 0 \,.$$

◇

On the contrary, decorrelation does not generally imply independence. An important exception is in the case of Gaussian vectors, discussed below.

C.2 Gaussian Vectors

Definition C.5 *A random vector* Z *of dimension* r *is a* Gaussian vector *if the characteristic function of* Z *is of the form*

$$\Phi_{\mathrm{Z}}(\theta) = \exp(i\theta^\top m_{\mathrm{Z}} - \frac{1}{2}\theta^\top \Lambda_{\mathrm{Z}}\theta) \,, \ \ \forall \theta \in \mathbb{R}^r \,, \tag{C.7}$$

where m_{Z} *is a vector of dimension* r *and* Λ_{Z} *a symmetric positive matrix of dimension* r.

A k-uple *of vectors* $(\mathrm{Z}_1, \ldots, \mathrm{Z}_k)$ *(of dimensions* r_1, ..., r_k*) is* Gaussian *if the vector* $(\mathrm{Z}_1^\top, \ldots, \mathrm{Z}_k^\top)^\top$ *(of dimension* $r_1 + \cdots + r_k$*) is Gaussian.*

Proposition C.6 *If* Z *is a Gaussian vector, then* Z *has moments of all orders and*

- *the vector* m_{Z} *in* (C.7) *coincides with the expectation of* Z,

$$m_{\mathrm{Z}} = \overline{\mathrm{Z}} = \mathbb{E}(\mathrm{Z}) ,$$

- *the symmetric positive matrix* Λ_{Z} *in* (C.7) *coincides with the dispersion of* Z,

$$\Lambda_{\mathrm{Z}} = \mathsf{D}(\mathrm{Z}) = \mathbb{E}\big[(\mathrm{Z} - \overline{\mathrm{Z}})(\mathrm{Z} - \overline{\mathrm{Z}})^{\top}\big] .$$

Proposition C.7 *An affine transformation of a Gaussian vector is a Gaussian vector.*

Proof Let Z be a Gaussian vector of dimension r, A be an $r \times r$ square matrix and b a vector of dimension r. We have that:

$$\begin{aligned}
\Phi_{\mathrm{AZ+b}}(\theta) &= \mathbb{E}\big[\exp\big(i\theta^{\top}(\mathrm{AZ} + \mathrm{b})\big)\big] \\
&= e^{i\theta^{\top}\mathrm{b}}\mathbb{E}\big[\exp(i(\mathrm{A}^{\top}\theta)^{\top}\mathrm{Z})\big] \\
&= e^{i\theta^{\top}\mathrm{b}}\Phi_{\mathrm{Z}}(\mathrm{A}^{\top}\theta) \\
&= \exp(i\theta^{\top}\mathrm{b} + i\theta^{\top}\mathrm{A}m_{\mathrm{Z}} - \frac{1}{2}\theta^{\top}\mathrm{A}\Lambda_{\mathrm{Z}}\mathrm{A}^{\top}\theta) \text{ by (C.7)} \\
&= \exp\left[i\theta^{\top}(\mathrm{b} + \mathrm{A}m_{\mathrm{Z}}) - \frac{1}{2}\theta^{\top}(\mathrm{A}\Lambda_{\mathrm{Z}}\mathrm{A}^{\top})\theta\right] .
\end{aligned}$$

Therefore, by (C.7) and Proposition C.6, $\mathrm{AZ} + \mathrm{b}$ is a Gaussian vector with mean $\mathrm{A}m_{\mathrm{Z}} + \mathrm{b}$ and dispersion matrix $\mathrm{A}\Lambda_{\mathrm{Z}}\mathrm{A}^{\top}$. □

Proposition C.8 *Let* $\mathrm{Z} = (\mathrm{X}, \mathrm{Y})$ *be a Gaussian couple. If* X *and* Y *are decorrelated random vectors, that is, if* $\mathsf{cov}(\mathrm{X}, \mathrm{Y}) = 0$, *then* X *and* Y *are independent (Gaussian) random vectors.*

Proof As $\mathbb{E}(\mathrm{Z}) = \begin{pmatrix}\overline{\mathrm{X}}\\ \overline{\mathrm{Y}}\end{pmatrix}$, and $\mathsf{D}(\mathrm{Z}) = \begin{pmatrix}\mathsf{D}(\mathrm{X}) & \mathsf{cov}(\mathrm{X},\mathrm{Y})\\ \mathsf{cov}(\mathrm{Y},\mathrm{X}) & \mathsf{D}(\mathrm{Y})\end{pmatrix}$, we have that

$$\begin{aligned}
\Phi_{\mathrm{Z}}(\theta) &= \Phi_{(\mathrm{X},\mathrm{Y})}(\theta_{\mathrm{X}}, \theta_{\mathrm{Y}}) \\
&= \exp\left(i\theta^{\top}\overline{\mathrm{Z}} - \frac{1}{2}\theta^{\top}\mathsf{D}(\mathrm{Z})\theta\right) \text{ by (C.7) and Proposition C.6} \\
&= \exp\left(i\theta^{\top}\begin{pmatrix}\overline{\mathrm{X}}\\ \overline{\mathrm{Y}}\end{pmatrix} - \frac{1}{2}\theta^{\top}\begin{pmatrix}\mathsf{D}(\mathrm{X}) & \mathsf{cov}(\mathrm{X},\mathrm{Y})\\ \mathsf{cov}(\mathrm{Y},\mathrm{X}) & \mathsf{D}(\mathrm{Y})\end{pmatrix}\theta\right) \\
&= \exp\left(i\theta^{\top}\begin{pmatrix}\overline{\mathrm{X}}\\ \overline{\mathrm{Y}}\end{pmatrix} - \frac{1}{2}\theta^{\top}\begin{pmatrix}\mathsf{D}(\mathrm{X}) & 0\\ 0 & \mathsf{D}(\mathrm{Y})\end{pmatrix}\theta\right) \text{ by decorrelation} \\
&= \exp\left(i\theta_{\mathrm{X}}^{\top}\overline{\mathrm{X}} - \frac{1}{2}\theta_{\mathrm{X}}^{\top}\mathsf{D}(\mathrm{X})\theta_{\mathrm{X}}\right)\exp\left(i\theta_{\mathrm{Y}}^{\top}\overline{\mathrm{Y}} - \frac{1}{2}\theta_{\mathrm{Y}}^{\top}\mathsf{D}(\mathrm{Y})\theta_{\mathrm{Y}}\right) \\
&= \Phi_{\mathrm{X}}(\theta_{\mathrm{X}})\Phi_{\mathrm{Y}}(\theta_{\mathrm{Y}}) ,
\end{aligned}$$

and we conclude thanks to Proposition C.2. □

The previous proof can easily be adapted to establish the following proposition.

Proposition C.9 *If* $\mathtt{X}$ *and* $\mathtt{Y}$ *are independent Gaussian vectors, the couple* $(\mathtt{X}, \mathtt{Y})$ *is a Gaussian couple.*

The following lemma is a result of orthogonal projection.

Lemma C.10 *Let* $\mathtt{X}$ *and* $\mathtt{Y}$ *two random vectors (with respective dimensions* $r_{\mathtt{X}}$ *and* $r_{\mathtt{Y}}$*). There exists a unique random vector* $\mathtt{Z}$*, denoted* $\pi_{\mathtt{Y}}(\mathtt{X})$*, with the same dimension as that of* $\mathtt{X}$*, such that*

1. *the vector* $\mathtt{Z} - \overline{\mathtt{X}}$ *is a linear expression of* $\mathtt{Y} - \overline{\mathtt{Y}}$*;*
2. *the random vector* $\mathtt{X} - \mathtt{Z}$ *is orthogonal to the random vector* $\mathtt{Y}$*, in the sense that* $\mathsf{cov}(\mathtt{X} - \mathtt{Z}, \mathtt{Y}) = 0$.

Proposition C.11 *Let* $(\mathtt{X}, \mathtt{Y})$ *a Gaussian couple. Then,* $\mathtt{X} - \pi_{\mathtt{Y}}(\mathtt{X})$ *is independent of* $\mathtt{Y}$*, and the conditional distribution of* $\mathtt{X}$ *knowing* $\mathtt{Y}$ *is Gaussian with mean* $\mathbb{E}\big[\mathtt{X} \mid \mathtt{Y}\big] = \pi_{\mathtt{Y}}(\mathtt{X})$ *and dispersion* $\mathsf{D}\left(\mathtt{X} - \pi_{\mathtt{Y}}(\mathtt{X})\right)$.

Proof Since $\pi_{\mathtt{Y}}(\mathtt{X})$ is an affine expression in $\mathtt{Y}$, the couple $(\mathtt{X}-\pi_{\mathtt{Y}}(\mathtt{X}), \mathtt{Y})$ is Gaussian by Proposition C.7. As $\mathtt{X} - \pi_{\mathtt{Y}}(\mathtt{X})$ is decorrelated of $\mathtt{Y}$, it is independent of $\mathtt{Y}$ by Proposition C.8. Moreover, we have that

$$
\begin{aligned}
&\mathbb{E}\left[\exp\left(i\theta_{\mathtt{X}}^{\top}\mathtt{X} + i\theta_{\mathtt{Y}}^{\top}\mathtt{Y}\right)\right] \\
&= \mathbb{E}\left[\exp\left(i\theta_{\mathtt{X}}^{\top}(\mathtt{X} - \pi_{\mathtt{Y}}(\mathtt{X})) + i\theta_{\mathtt{X}}^{\top}\pi_{\mathtt{Y}}(\mathtt{X}) + i\theta_{\mathtt{Y}}^{\top}\mathtt{Y}\right)\right] \\
&= \mathbb{E}\left[\exp\left(i\theta_{\mathtt{X}}^{\top}(\mathtt{X} - \pi_{\mathtt{Y}}(\mathtt{X})\right)\right]\mathbb{E}\left[\exp\left(i\theta_{\mathtt{X}}^{\top}\pi_{\mathtt{Y}}(\mathtt{X}) + i\theta_{\mathtt{Y}}^{\top}\mathtt{Y}\right)\right] \\
&\quad \text{by independence of } \mathtt{X} - \pi_{\mathtt{Y}}(\mathtt{X}) \text{ and } \mathtt{Y} \\
&= \exp\left(-\frac{1}{2}\theta_{\mathtt{X}}^{\top}\,\mathsf{D}(\mathtt{X} - \pi_{\mathtt{Y}}(\mathtt{X}))\theta_{\mathtt{X}}\right)\mathbb{E}\left[\exp\left(i\theta_{\mathtt{X}}^{\top}\pi_{\mathtt{Y}}(\mathtt{X}) + i\theta_{\mathtt{Y}}^{\top}\mathtt{Y}\right)\right] \\
&\quad \text{because } \mathtt{X} - \pi_{\mathtt{Y}}(\mathtt{X}) \text{ is centered Gaussian} \\
&= \mathbb{E}\left[\left(\exp(-\frac{1}{2}\theta_{\mathtt{X}}^{\top}\,\mathsf{D}(\mathtt{X} - \pi_{\mathtt{Y}}(\mathtt{X}))\theta_{\mathtt{X}}\right)\exp\left(i\theta_{\mathtt{X}}^{\top}\pi_{\mathtt{Y}}(\mathtt{X})\right)\exp\left(i\theta_{\mathtt{Y}}^{\top}\mathtt{Y}\right)\right].
\end{aligned}
$$

By definition of the conditional expectation, we obtain:

$$
\mathbb{E}\big[\exp\left(i\theta_{\mathtt{X}}^{\top}\mathtt{X}\right) \mid \mathtt{Y}\big] = \exp\left(-\frac{1}{2}\theta_{\mathtt{X}}^{\top}\,\mathsf{D}(\mathtt{X} - \pi_{\mathtt{Y}}(\mathtt{X}))\theta_{\mathtt{X}}\right)\exp\left(i\theta_{\mathtt{X}}^{\top}\pi_{\mathtt{Y}}(\mathtt{X})\right).
$$

By comparison with (C.7), and by Proposition C.2, we deduce that the conditional distribution of $\mathtt{X}$ knowing $\mathtt{Y}$ is Gaussian, with mean $\mathbb{E}\big[\mathtt{X} \mid \mathtt{Y}\big] = \pi_{\mathtt{Y}}(\mathtt{X})$ and dispersion $\mathsf{D}\left(\mathtt{X} - \pi_{\mathtt{Y}}(\mathtt{X})\right)$. □

The following lemmas are direct applications of Proposition C.11 and of properties of the orthogonal projection.

Lemma C.12 *Let* $(\mathtt{X}, \mathtt{Y}, \mathtt{Z})$ *be a Gaussian triple such that* $\mathtt{Y}$ *and* $\mathtt{Z}$ *are decorrelated. Then, we have that*

$$\mathbb{E}\big[\mathtt{X} \mid \mathtt{Y}, \mathtt{Z}\big] = \mathbb{E}\big[\mathtt{X} \mid \mathtt{Y}\big] + \mathtt{L}(\mathtt{Z} - \overline{\mathtt{Z}}) \,, \tag{C.8}$$

where the matrix $\mathtt{L}$ *satisfies* $\mathsf{cov}(\mathtt{X}, \mathtt{Z}) = \mathtt{L}\,\mathsf{D}(\mathtt{Z})$.

Lemma C.13 *Let* $(\mathtt{X}, \mathtt{Y})$ *be a Gaussian couple such that* $\mathsf{D}(\mathtt{Y})$ *is a nondegenerate matrix. We have that:*

$$\begin{cases} \mathbb{E}\big[\mathtt{X} \mid \mathtt{Y}\big] & = \overline{\mathtt{X}} + \mathsf{cov}(\mathtt{X}, \mathtt{Y})\,\mathsf{D}(\mathtt{Y})^{-1}(\mathtt{Y} - \overline{\mathtt{Y}}) \\ \mathsf{D}\left(\mathtt{X} - \mathbb{E}\big[\mathtt{X} \mid \mathtt{Y}\big]\right) & = \mathsf{D}(\mathtt{X}) - \mathsf{cov}(\mathtt{X}, \mathtt{Y})\,\mathsf{D}(\mathtt{Y})^{-1}\,\mathsf{cov}(\mathtt{X}, \mathtt{Y})^{\top} \,. \end{cases} \tag{C.9}$$

Appendix D
Bode Diagrams

This appendix completes § 3.7.

The *Nyquist diagram* of a transfer function $\mathtt{H}(s)$ is the curve $\omega \mapsto \mathtt{H}(i\omega)$ represented in the complex plane and parameterized by the increasing pulsations, ω varying from 0 to $+\infty$. It can also be represented by two curves parameterized by ω, the amplitude curve $|\mathtt{H}(i\omega)|$ and the phase curve $\mathrm{Arg}\big(\mathtt{H}(i\omega)\big)$.

Remark D.1 The amplitude curve generally displays the amplitude in dB (decibels), that is, as $20 \log_{10} |\mathtt{H}(i\omega)|$, whereas the abscissa axis ω is graduated in logarithmic coordinates.

Definition D.2 *The amplitude and phase curves of* $\mathtt{H}(i\omega)$ *are respectively called* amplitude Bode diagram *or* amplitude Bode plot *of* $\mathtt{H}(i\omega)$ *and* phase Bode diagram *or* phase Bode plot *of* $\mathtt{H}(i\omega)$.

Example D.3

- Figure D.1 displays (in the case $T = 1$ in (D.1) below) the amplitude and phase diagrams of a first-order system with transfer function:

$$\mathtt{H}(s) = \frac{1}{1 + Ts}\ , \quad T > 0\ . \tag{D.1}$$

- Figure D.2 displays (in the case $\omega_n = 1$ in (D.2) below) the amplitude and phase diagrams of a second-order system with transfer function:

$$\mathtt{H}(s) = \frac{\omega_n^2}{s^2 + 2\zeta\omega_n s + \omega_n^2}\ . \tag{D.2}$$

△

More often, we are interested in *asymptotic Bode diagrams* displayed in Figs. D.3 and D.4. Their interest stems from the fact that it is rather easy to obtain the asymptotic

B. d'Andréa-Novel and M. De Lara, *Control Theory for Engineers*,
DOI: 10.1007/978-3-642-34324-7, © Springer-Verlag Berlin Heidelberg 2013

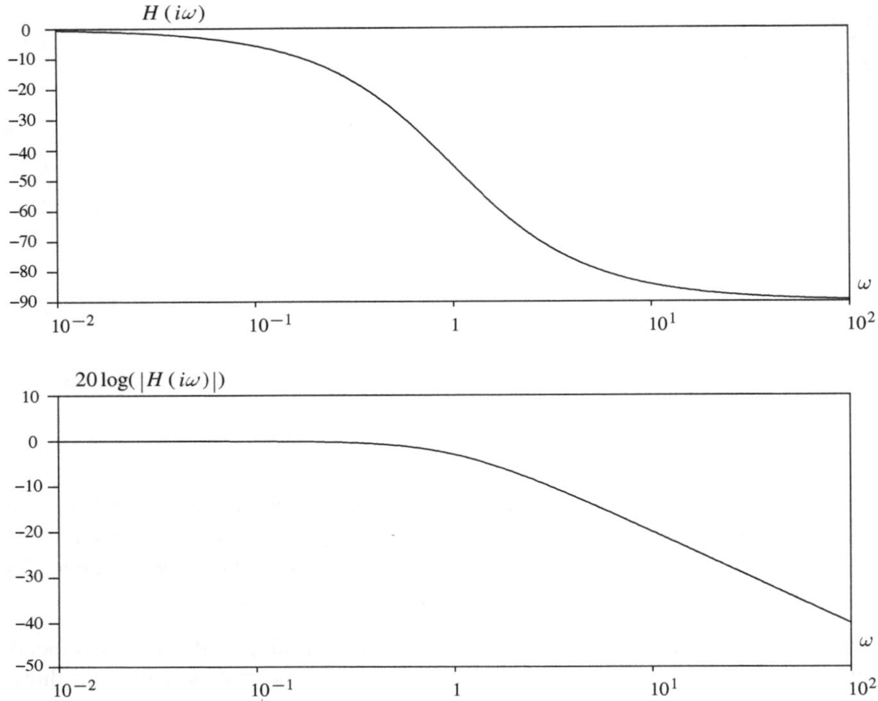

Fig. D.1 Amplitude and phase bode diagrams of a first-order system (D.1) (with $T = 1$)

Bode diagrams of any transfer function $\mathrm{H}(s)$, knowing the asymptotic Bode diagrams of first and second-order systems. In fact, $\mathrm{H}(s)$ can be written as the product of transfer functions $\mathrm{H}_i(s)$ of first and second order:

$$\mathrm{H}(s) = \mathrm{H}_1(s) \times \cdots \times \mathrm{H}_r(s) \ .$$

But, since $\log H = \log \mathrm{H}_1 + \cdots + \log \mathrm{H}_r$, the logarithmic diagrams of $\mathrm{H}(s)$ can be obtained by a simple addition of the asymptotic diagrams of the $\mathrm{H}_i(s)$ with breakpoints at each pulsation ω_n corresponding to a change of slope in the amplitude curve:

- -20 dB/decade for a real pole, viz. $\mathrm{H}_i(s) = \dfrac{1}{1 + s/\omega_n}$;
- $+20$ dB/decade for a real zero, viz. $\mathrm{H}_i(s) = 1 + s/\omega_n$;
- -40 dB/decade for complex conjugate poles, viz.

$$\mathrm{H}_i(s) = \frac{1}{1 + 2\zeta s/\omega_n + s^2/\omega_n^2} \ ;$$

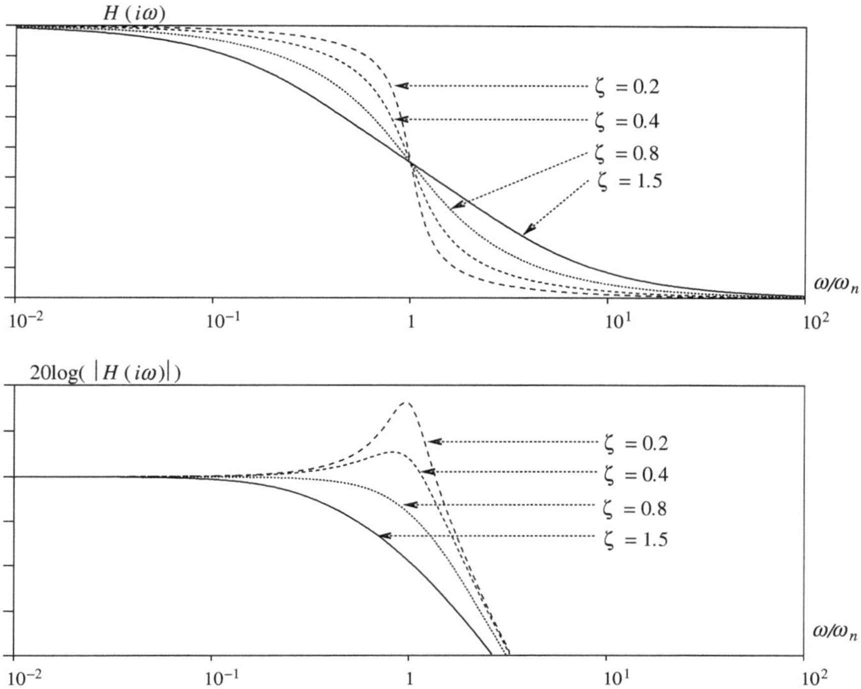

Fig. D.2 Amplitude and phase bode diagrams of a second-order system (D.2) (with $\omega_n = 1$)

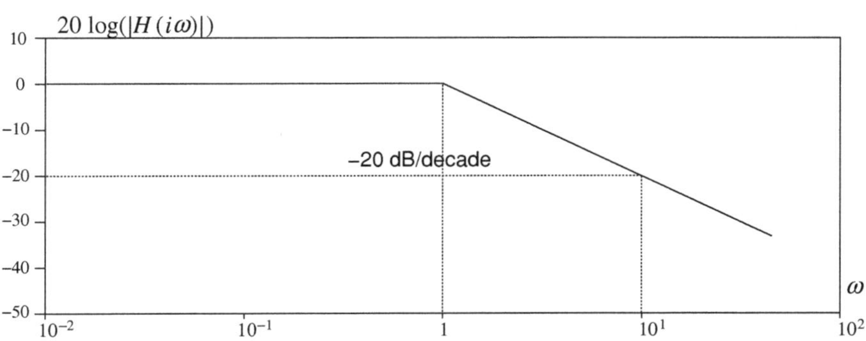

Fig. D.3 Asymptotic amplitude bode diagram of a first-order system (D.1) (with $T = 1$)

- +40 dB/decade for complex conjugate zeros, viz.

$$\mathrm{H}_i(s) = 1 + 2\zeta s/\omega_n + s^2/\omega_n^2 \,.$$

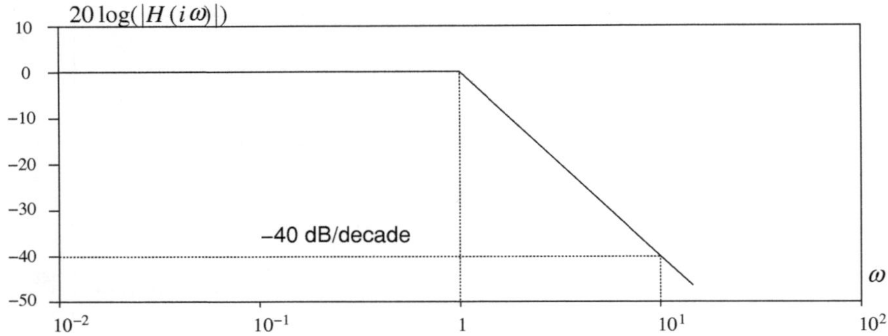

Fig. D.4 Asymptotic amplitude bode diagram of a second-order system (D.2) (with $\omega_n = 1$)

From Bode diagrams, one can also define the *gain and phase margins*, as illustrated in Figure D.5.

The gain margin γ is equal to $(1 - \gamma_0)$, γ_0 being the gain corresponding to the phase of 180°, namely $20\log_{10}\gamma$ (dB). The phase margin ϕ is equal to (180° $-\phi_1$), where ϕ_1 is the phase corresponding to the unit gain (namely 0 dB).

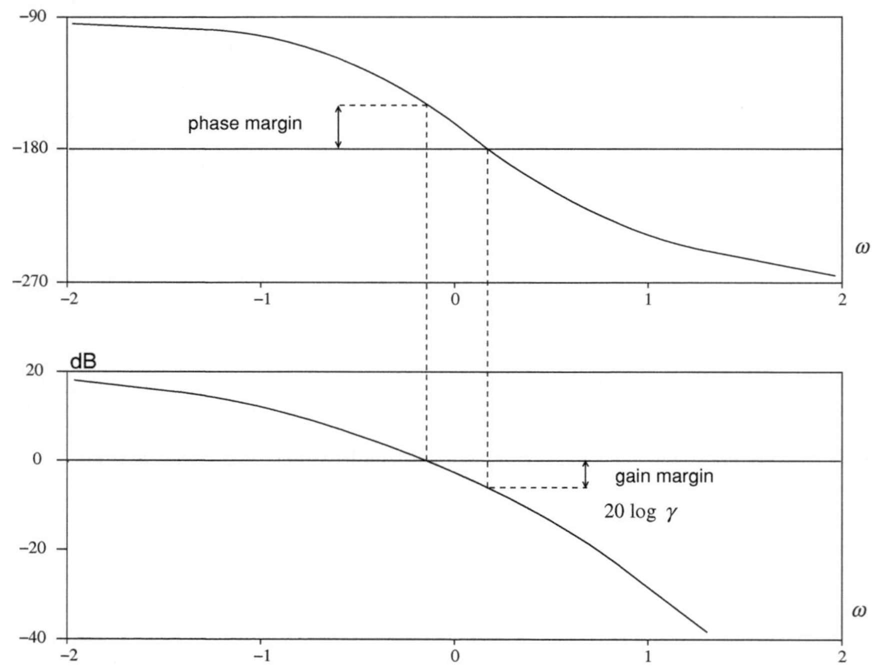

Fig. D.5 Gain and phase margins

References

1. R. Abraham, J.E. Marsden, T. Ratiu, *Manifolds, Tensor Analysis, and Applications* (Springer, New York, 1988)
2. J. Ackermann, *Sampled-Data Control Systems* (Springer, Berlin, 1985)
3. D.V. Anosov, V.I. Arnold (eds.), *Dynamical Systems I* (Springer, Berlin, 1988)
4. P. Appell, *Traité de mécanique rationnelle* (Gauthier-Villars, Paris, 1903)
5. V.I. Arnold, *Ordinary Differential Equations* (The MIT Press, Cambridge, 1988)
6. K. Åström, B. Wittenmark, *Computer Controlled Systems: Theory and Design*, Information and Systems Sciences Series (Prentice-Hall, New Jersey, 1984)
7. P. Atkins, L. Jones, *Chemistry* (W. H. Freeman and Company, New York, 1997)
8. G. Bastin, D. Dochain, *On-Line Estimation and Adaptive Control of Bioreactors* (Elsevier, Amsterdam, 1990)
9. E. Beltrami, *Mathematics for Dynamic Modeling* (Academic Press, Orlando, 1987)
10. W.M. Boothby, *An Introduction to Differentiable Manifolds and Riemannian Geometry*, 2nd edn. (Academic, New York, 2003)
11. L. Breiman, *Probability Classics in Applied Mathematics*, 2nd edn. (siam, Philadelphia, 1993)
12. R. Brockett, *Finite Dimensional Linear Systems* (Wiley, New York, 1973)
13. B.S.C. Canudas de Wit, G. Bastin (eds.), *Theory of Robot Control*, 2nd edn. (Springer, 1997)
14. P. Caines, D. Maynes, On the discrete time matrix Riccati equation of optimal control. Int. J. Control **12**(5), 785–794 (1970)
15. P. Caines, D. Maynes, On the discrete time matrix Riccati equation of optimal control-a correction. Internat. J. Control **14**(1), 205–207 (1971)
16. S. Campbell, J.-P. Chancelier, R. Nikoukhah, *Modeling and Simulation in Scilab/Scicos with ScicosLab 4.4*, 2nd edn. (Springer, New York, 2010)
17. G. Campion, B. d'Andréa-Novel, G. Bastin, in *Controllability and State Feedback Stabilization of Nonholonomic Mechanical Systems*, ed. by C. de Wit. Advanced Robot Control. Lecture Notes in Control and Information Sciences, vol. 162, pp. 106–124 1990
18. H. Cartan, *Cours de calcul différentiel* (Hermann, Paris, 1979)
19. H. Caswell, *Matrix Population Models* (Sinauer Associates Inc., Sunderland, 1989)
20. A.J. Chorin, J.E. Marsden, *A Mathematical Introduction to Fluid Mechanics* (Springer, New York, 1979)
21. M. Cibrario. Sur la commande d'une classe de réacteurs chimiques instables, PhD thesis, École nationale supérieure des mines de Paris, 1992
22. P. Coiffet, La robotique, principes et applications (Hermes, Paris, 1986)
23. J.-M. Coron, *Control and Nonlinearity, Volume 136 of Mathematical Surveys and Monographs* (American Mathematical Society, Providence, 2007)

B. d'Andréa-Novel and M. De Lara, *Control Theory for Engineers*,
DOI: 10.1007/978-3-642-34324-7, © Springer-Verlag Berlin Heidelberg 2013

24. J. Craig. Introduction to Robotics: Mechanics and Control. Pearson - International Edition, 3 edition, 2003
25. B. d'Andréa-Novel, Commande non linéaire des robots (Hermes, Paris, 1988)
26. B. d'Andréa-Novel, L. Praly, Playing with the regulation zeros in the stabilization of a double inverted pendulum. Kybernetica **27**(3), 213–224 (1991)
27. M. De Lara, L. Doyen, *Sustainable Management of Natural Resources* (Mathematical Models and Methods. Springer-Verlag, Berlin, 2008)
28. M. Demazure, *Catastrophes et bifurcations* (Ellipses Éditions Marketing, Paris, 1989)
29. S. Dreyfus, Richard Bellman on the birth of dynamic programming. Oper. Res. **50**(1), 48–51 (2002)
30. P. Faurre, M. Robin, *Éléments d'automatique* (Dunod, Paris, 1984)
31. W. Feller, *An Introduction to Probability Theory and its Applications, volume 1*, 3rd edn. (Wiley, New York, 1968)
32. R. Feynman, R. Leighton, M. Sands, *The Feynman Lectures on Physics*, 2nd edn. (Addison-Wesley, Boston, 1963)
33. B. Friedland, *Control System Design* (Mac Graw-Hill, New York, 1986)
34. F.R. Gantmacher, The Theory of Matrices, vol. 1 (AMS Chelsea Publishing, Providence, 1959, 1960, 1977)
35. F.R. Gantmacher, The Theory of Matrices, vol. 2 (AMS Chelsea Publishing, Providence, 1959, 1960, 1977)
36. I. Gohberg, P. Lancaster, L. Rodman, *Matrix Polynomials* (Academic Press, New York, 1982)
37. J. S. H. Goldstein, C.P. Poole. Classical Mechanics. Pearson - International Edition, 3 edition, 2001.
38. J.K. Hale, *Ordinary Differential Equations* (Wiley-Interscience, New York, 1969)
39. M. Hervé, *Les fonctions analytiques* (PUF, Paris, 1982)
40. A. Hurwitz, Uber die bedingungen, unter welchen einer gleichung nur wurzeln mit negativen reelen teilen besitzt. Math. Ann. **146**, 273–284 (1895)
41. A.H. Jazwinski, *Stochastic Processes and Filtering Theory* (Academic Press, San Diego, 1970)
42. R.W. Howell, J.H. Mathews, *Complex Analysis for Mathematics and Engineering*, 5th edn. (Jones and Bartlett Publishers Inc., Boston, 2006)
43. E.I. Jury, *Theory and Application of the z-Transform Method* (Wiley, New York, 1964)
44. T. Kailath, *Linear Systems*, Information and Systems Sciences Series (Prentice-Hall, New Jersey, 1980)
45. W. Khalil, E. Dombre, *Modélisation et commande des robots* (Hermes, Paris, 1988)
46. J. LaSalle, S. Lefschetz, *Stability by Lyapunov's Direct Method* (Academic Press, New York, 1961)
47. A.J.G. MacFarlane, *Frequency Response Methods in Control Systems* (IEEE Press, New York, 1979)
48. M. Mesterton-Gibbons, *A Concrete Approach To Mathematical Modelling* (Addison Wesley, Redwood City, California, 1989)
49. W. Michiels, S.-I. Niculescu, Stability and Stabilization of Time-Delay Systems. An Eigenvalue-Based Approach, vol. 12 (SIAM, Philadelphia, 2007)
50. J. Palis, W. de Melo, *Geometric Theory of Dynamical Systems* (Springer, Berlin, 1982)
51. L. Pernebo, An algebraic theory for the design of controllers for linear multivariable systems. part I: structure matrices and feedforward design. IEEE Trans. Automat. Control **26**(1), 171–182 (1981a)
52. L. Pernebo, An algebraic theory for the design of controllers for linear multivariable systems. part II: feedback realizations and feedback design. IEEE Trans. Automat. Control **26**(1), 183–194 (1981b)
53. J.F. Pommaret, A. Quadrat, Generalized Bezout identity. Appl. Algebra Eng. Commun. Comput. **9**(2), 91–116 (1998)
54. L. Praly, Closed loop transfer functions and internal stability. in *Proceedings of the 20th CDC*, San Diego, California, 1981, pp. 298–300
55. H.H. Rosenbrock, *State-Space and Multivariable Theory* (Wiley, New York, 1970)

56. H.H. Rosenbrock, *Computer Aided Control System Design* (Academic Press, New York, 1974)
57. N. Rouche, J. Mawhin, *Équations différentielles ordinaires*, vol. 2 (Masson, Paris, 1973)
58. E.J. Routh, *A Treatise on the Stability of a Given State of Motion* (Macmillan & Co., London, 1877)
59. D. Russell, in *Mathematical Models for the Elastic Beam and Their Control Theoretic Implications*, ed. by M. G. H. Brezis, F. Kappel. Semi-Groups, Theory and Applications (Longman Scientific and Technical, UK, 1985), pp. 177–216
60. L. Schwartz, *Méthodes mathématiques pour les sciences physiques* (Hermann, Paris, 1969)
61. R.J. Schwartz, B. Friedland, *Linear Systems* (Mac Graw-Hill, New York, 1965)
62. J. Slotine, W. Li, Adaptive manipulator control : a case study. IEEE Trans. Automat. Control **33**(11), 995–1003 (1988)
63. E.D. Sontag, *Mathematical Control Theory* (Springer, New York, 1990)
64. M. Vidyasagar, *Nonlinear Systems Analysis* (Prentice Hall, New Jersey, 1978)
65. M. Vukobratović, R. Stojić, *Modern Aircraft Flight Control* (Springer, Berlin, 1988)
66. J. Willems, Least squares stationary optimal control and the algebraic Riccati equation. IEEE Trans. Automat. Control **16**(6), 621–634 (1971)
67. J. Willems. Les espaces presque-invariants. In Outils et modèles mathématiques pour l'automatique, l'analyse de systèmes et le traitement du signal, vol. 3 (Éditions du CNRS, Paris, 1983), pp. 177–219
68. P.Y. Willems, *Introduction à la mécanique* (Masson, Paris, 1979)
69. W.A. Wolovich, *Linear Multivariable Systems* (Springer, New York, 1974)
70. W.M. Wonham, *Linear Multivariable Control: A Geometric Approach* (Springer, New York, 1979)
71. D. Youla, H. Jabr, J. Bongiorno, Modern Wiener Hopf design of optimal controllers, part II: The multivariable case. IEEE Trans. Automat. Control **21**, 319–338 (1976)

Index

B. d'Andréa-Novel and M. De Lara, *Control Theory for Engineers*,
DOI: 10.1007/978-3-642-34324-7, © Springer-Verlag Berlin Heidelberg 2013

MIX
Papier aus verantwortungsvollen Quellen
Paper from responsible sources
FSC® C105338

If you have any concerns about our products,
you can contact us on
ProductSafety@springernature.com

In case Publisher is established outside the EU,
the EU authorized representative is:
Springer Nature Customer Service Center GmbH
Europaplatz 3, 69115 Heidelberg, Germany

Printed by Libri Plureos GmbH
in Hamburg, Germany